A Textbook of Environmental Studies

Published by:
Saurabh Publishing House

A Textbook of Environmental Studies

Dr. Ashok Dawan

Saurabh
Publishing House
4735/22, Prakash Deep Building,
Ansari Road, Daryaganj,
New Delhi - 110002

Published by
SAURABH PUBLISHING HOUSE

Distributed by:
Lotus Press Publishers & Distributors
Unit No. 220, Second Floor, 4735/22, Prakash Deep Building,
Ansari Road, Darya Ganj, New Delhi-110002
Ph.: 011-23280047, 32903912, 9811838000
E-mail: lotus_press@sify.com, www.lotuspress.co.in

Saurabh Publishing House is an imprint of
Lotus Press Publishers & Distributors

A Textbook of Enviromental Studies

ISBN: 978-81-89005-22-1

Printed at: Concept Imprint, Delhi

PREFACE

Environmental issues are hardly new subjects of discussion, deliberation and action with ever increasing commercialisation and urbanisation making their presence felt in a continually globalising world, environmental studies have only stressed their relevance more in society. Even as reams of newspaper, public and private forums of debate, internet websites and other media forms devote attention to the challenges confronting humanity through environmental degradation, the extent of human activities in the world has reached such a level as to bringing up a new environmental problem almost every single day. In such a scenario, it becomes crucial to emphasise upon the need for environmental studies, what it brings into consideration, and how it can redeem a situation which threatens the very fabric of human life.

This book, written as an introductory manual, brings within its purview the subject of environmental studies, its principles, processes, practices concepts, and the import it carries in a world which can no longer turn a blind eye to the environmental problems which face it. It guides the readers through a comprehensive content which focuses upon both traditional and contemporary problems of the environment, the local, national and international environmental policies, the role of the common man, and the destruction and degradation of the environment. In the field's concerns, and how they expand the scope of the insight serve resourceful for the readers.

Author

CONTENTS

	Preface	*v*
1.	Understanding Our Environment	1
2.	The Ecosystem	35
3.	Natural Resources	71
4.	Food Resources	91
5.	State of Global Agriculture	105
6.	Freshwater Resources	131
7.	Global Forest Wealth	153
8.	Energy and Environment	185
9.	Biodiversity and Conservation	223
10.	Natural Disasters	253
11.	Environmental Pollution	267
12.	Legal Framework for Environmental Protection	287
	Bibliography	307
	Index	309

1

UNDERSTANDING OUR ENVIRONMENT

NATURE AND SCOPE OF ENVIRONMENTAL STUDIES

Environmental studies is the systematic study of human interaction with their environment. It is a broad field of study that includes the natural environment, built environments, social environments, organisational environments, and the sets of relationships between them. Several schools of environmental studies exist throughout the world. The demand and creation of such schools is likely associated with the growing environmental concerns of global climate change, biodiversity loss, pollution and environmental sustainability. Universities are also adding such schools as institutes within their programmes.

Current environmental problems have evolved into a complex set of interdisciplinary issues involving ecological, political, economic, social, as well as physical and biological considerations. Modern environmental studies must include the study of the urban environment as well as the natural environment. Erratic weather cycles, increasing temperatures, flooded cities and depleting natural resources. With increasing unorganised and uncontrolled development, environmental studies has become an important course of study for students today. In recent years environmental studies has emerged as an important discipline that can offer solutions to many environmental problems. Several scientific disciplines like biochemistry, physics, mathematics, biotechnology, chemistry, botany, toxicology, remote sensing and engineering have an interface with environmental science.

Environmental scientists monitor the quality of the environment, interpret the impact of human actions on terrestrial and aquatic ecosystems, and develop strategies for restoring ecosystems. In addition, environmental scientists help planners develop and construct buildings, transportation corridors, and utilities that protect water resources and reflect efficient and beneficial land use.

Environmental studies encompasses issues such as climate change, conservation, biodiversity, groundwater and soil contamination, use of natural resources, waste management, sustainable development, air pollution and noise pollution. While the environment has been studied for at least as long as there has been science, the recent interest in putting the pieces of understanding together to study environmental systems has come alive as a substantive, active field of scientific investigation starting in the 1960s and 1970s. This has been driven by the need for a large multi-disciplined team to analyse complex environmental problems, the arrival of substantive environmental laws requiring specific environmental protocols of investigation, and growing public awareness of a need for action in addressing environmental problems.

The magnitude and complexity of environmental problems are creating a growing need for scientists with rigorous, interdisciplinary training in environmental science. The majority of environmental scientists are employed in governmental positions, but the job market in the private sector is expected to see the most growth because of public policy demand for new rules and regulations. Job growth for environmental scientists should be strongest in private-sector consulting firms. Growth in employment of environmental scientists will be spurred largely by the increasing demands placed on the environment and water resources by population growth. Further demand should result from the need to comply with complex environmental laws and regulations, particularly those regarding groundwater decontamination, clean air, and flood control. Employment of environmental scientists is expected to increase by 25 per cent between 2006 and 2016, much faster than the average for all occupations.

Closely linked with ecology, environmental studies is primarily concerned with how humanity affects and in turn is affected by other living organisms and the non-living physical environment. Students get an exposure to varied subjects such as biodiversity, microbiology, pollution control methods, disaster management, environment impact studies and conservation of natural resources. In India, these basic disciplines of science have taken strong roots over a period of several decades but environmental studies is now an emerging field through interdisciplinary collaboration.

Environmental studies is offered at the undergraduate level at Delhi University, University of Mysore, Pune University and Bharathiar University—Coimbatore. Apart from theory and practical sessions, the course also involves field trips and surveys. After graduation, interested students can also pursue their masters and research programme in the same field. Comprehensive M.Sc. programmes are offered at Andhra University, Annamalai University Tamil Nadu, Centre for Ecological Research at the Indian Institute of Science in Bangalore, Centre for Environmental Studies at the Indian Institute of Technology in Mumbai, Indian Institute of Ecology and Environment in New Delhi, Jawaharlal Nehru University, Delhi University and Cochin University of Science and Technology. Like most masters programmes students have to pass an entrance examination and should have completed their graduation in a science stream. Increasing public awareness on the importance of conserving the environment has led many universities to adopt the programme into their curriculum.

Any course of development today requires the guidance of an environment specialist. Environmental studies has become an important part of urban planning including the construction of houses, sanitation, water management and waste disposal. Apart from the routine study and research options environmental studies students today can secure jobs in pollution control boards, public health laboratories, irrigation and agriculture department, chemical industries, research industries such as Central Drug Research Institute, Central Institute for Medicinal and Aromatic Plants, Industrial Toxicology Research Centre, National Environmental Engineering Research Institute, Environment Protection and Training Research Institute and Tata Energy Research Institute and students can also develop their own laboratories for analytical work in non-governmental sectors.

The Supreme Court judgment has made the study of environmental studies compulsory for all students from school to postgraduate level. This has also opened numerous teaching options for environmental studies students. Students also have avenues in wildlife conservation. The 'green science ' has also found its way into engineering programmes and institutions such as Andhra University College of Engineering, Delhi College of Engineering and L.D. College of Engineering, Gujarat University offer specialisation in environmental science engineering.

COMPONENTS OF NATURAL ENVIRONMENT

The natural environment, commonly referred to simply as the environment, is a term that comprises all living and non-living things that occur naturally on Earth or some part of it. This term includes a few key components:

1. *Complete landscape units* that function as natural systems without massive human intervention, including all plants, animals, rocks, etc. and natural phenomena that occur within their boundaries.
2. *Universal natural resources and physical phenomena* that lack clear-cut boundaries, such as air, water, and climate, as well as energy, radiation, electric charge, and magnetism, not originating from human activity.

Within the various uses of the word today, "nature" may refer to the general realm of various types of living plants and animals, and in some cases to the processes associated with inanimate objects—the way that particular types of things exist and change of their own accord, such as the weather and geology of the Earth, and the matter and energy of which all these things are composed. It is often taken to mean the "natural environment" or wilderness—wild animals, rocks, forest, beaches, and in general those things that have not been substantially altered by human intervention, or which persist despite human intervention.

The Earth

The Earth is the fifth largest planet in the solar system, third in order of distance from the Sun. It is the largest of its planetary system's terrestrial planets and the only place in the universe known to support life. The most prominent features of the earth's climate are its two large polar regions, two relatively narrow temperate zones, and a wide equatorial tropical to subtropical region. Precipitation patterns vary widely according to location, ranging from several metres of water per year to less than a millimetre. About 70 per cent of the surface is covered by saltwater oceans. The remainder consists of continents and islands, with the vast majority of the inhabited land in the Northern Hemisphere.

Earth has evolved through geological and biological processes that have left traces of the original conditions. The outer surface is divided into several tectonic plates that gradually migrate across the surface over geologic time spans, which at least several times have changed relatively quickly. The interior of the planet remains active, with a thick layer of molten Earth mantle and an iron-filled core that generates a magnetic field.

The atmospheric conditions have been significantly altered from the original conditions by the presence of life forms, which create an ecological balance that stabilises the surface conditions. Despite the wide regional variations in climate by latitude and other geographic factors, the long-term average global climate is quite stable during interglacial periods, and variations of a degree or two of average global temperature have historically had major effects on the ecological balance, and on the actual geography of the Earth.

Based on the available evidence, scientists have reconstructed detailed information about the planet's past. Earth is estimated to have formed approximately 4.55 billion years ago out of the solar nebula, along with the Sun and other planets. The Moon formed relatively soon afterwards.

Initially molten, the outer layer of the planet cooled, resulting in the solid crust. Outgassing and volcanic activity produced the primordial atmosphere.

Condensing water vapour, augmented by ice delivered by comets, produced the oceans. The highly energetic chemistry is believed to have produced a self-replicating molecule around 4 billion years ago.

Continents formed, then broke up and reformed as the surface of Earth reshaped itself over the course of hundreds of millions of years, occasionally combining to make a super-continent. Roughly 750 million years ago, the earliest known super-continent Rodinia, began to break apart. The continents later recombined to form Pannotia which broke apart about 540 million years ago, then finally Pangaea, which broke apart about 180 million years ago.

There is significant evidence, still being discussed among scientists, that a severe glacial action during the Neoproterozoic era covered much of the planet in a sheet of ice. This hypothesis has been termed the "Snowball Earth", and it is of particular interest as it precedes the Cambrian explosion in which multicellular life forms began to proliferate about 530-540 million years ago.

Since the Cambrian explosion there have been five distinctly identifiable mass extinctions. The last mass extinction occurred some 65 million years ago, when a meteorite collision probably triggered the extinction of the non-avian dinosaurs and other large reptiles, but spared small animals such as mammals, which then resembled shrews. Over the past 65 million years, mammalian life diversified.

Several million years ago, a species of small African ape gained the ability to stand upright. The subsequent advent of human life, and the development of agriculture and further civilisation allowed humans to affect the Earth more rapidly than any previous life form, affecting both the nature and quantity of other organisms as well as global climate.

The present era is classified as part of a mass extinction event, the Holocene extinction event, the fastest ever to have occurred. Some, such as E. O. Wilson of Harvard University, predict that human destruction of the biosphere could cause the extinction of one-half of all species in the next 100 years. The extent of the current extinction event is still being researched, debated and calculated by biologists.

Atmosphere

Earth's atmosphere is a layer of gases surrounding the planet Earth and retained by the Earth's gravity. It contains roughly 78 per cent nitrogen, 20.95 per cent oxygen, 0.93 per cent argon, 0.038 per cent carbon dioxide, trace amounts of other gases, and a variable amount of water vapour. This mixture of gases is commonly known as air. The atmosphere protects life on Earth by absorbing ultraviolet solar radiation and reducing temperature extremes between day and night.

There is no definite boundary between the atmosphere and outer space. It slowly becomes thinner and fades into space. Three quarters of the atmosphere's mass is within 11 km of the planetary surface. In the United States, people who travel above an altitude of 80.5 km are designated astronauts. An altitude of 120 km marks the boundary where atmospheric effects become noticeable during re-entry. The Karman line, at 100 km, is also frequently regarded as the boundary between atmosphere and outer space.

Temperature and Layers

The temperature of the Earth's atmosphere varies with altitude; the mathematical relationship between temperature and altitude varies among six different atmospheric layers:

1. *Exosphere:* From 500-1000 km up to 10,000 km, free-moving particles that may migrate into and out of the magnetosphere or the solar wind.
2. *Ionosphere:* is the part of the atmosphere that is ionised by solar radiation. It plays an important part in atmospheric electricity and forms the inner edge of the magnetosphere. It has practical importance because, among other functions, it influences radio propagation to distant places on the Earth. It is located in the thermosphere and is responsible for auroras.
3. *Thermosphere:* From 80-85 km to 640+ km, temperature increasing with height.
4. *Mesosphere:* The mesosphere extends from about 50 km to the range of 80 to 85 km, temperature decreasing with height. This is also where most meteors burn up when entering the atmosphere.
5. *Stratosphere:* From the Latin word "stratus" meaning a spreading out. The stratosphere extends from the troposphere's 7 to 17 km range to about 50 km. Temperature increases with height. The stratosphere contains the ozone layer, the part of the Earth's atmosphere which contains relatively high concentrations of ozone. "Relatively high" means a few parts per million—much higher than the concentrations in the lower atmosphere but still small compared to the main components of the atmosphere. It is mainly located in the lower portion of the stratosphere from approximately 15 to 35 km above Earth's surface, though the thickness varies seasonally and geographically.
6. *Troposphere:* The troposphere is the lowest layer of the atmosphere; it begins at the surface and extends to between 7 km at the poles and 17 km at the equator, with some variation due to weather

factors. The troposphere has a great deal of vertical mixing due to solar heating at the surface. This heating warms air masses, which makes them less dense so they rise. When an air mass rises the pressure upon it decreases so it expands, doing work against the opposing pressure of the surrounding air. To do work is to expend energy, so the temperature of the air mass decreases. As the temperature decreases, water vapour in the air mass may condense or solidify, releasing latent heat that further uplifts the air mass. This process determines the maximum rate of decline of temperature with height, called the adiabatic lapse rate. It contains roughly 80 per cent of the total mass of the atmosphere. 50 per cent of the total mass of the atmosphere is located in the lower 5 km of the troposphere.

The average temperature of the atmosphere at the surface of Earth is 15°C.

Pressure and Thickness

The average atmospheric pressure, at sea level, is about 101.3 kilopascals; total atmospheric mass is 5.1361×10^{18} kg. Atmospheric pressure is a direct result of the total weight of the air above the point at which the pressure is measured. This means that air pressure varies with location and time, because the amount of air above the earth varies with location and time.

Atmospheric pressure decreases with height, dropping by 50 per cent at an altitude of about 5.6 km. Equivalently, about 50 per cent of the total atmospheric mass is within the lowest 5.6 km. This pressure drop is approximately exponential, so that pressure decreases by approximately half every 5.6 km. However, because of changes in temperature throughout the atmospheric column, as well as the fact that the force of gravity begins to decrease at great altitudes, a single equation does not model atmospheric pressure through all altitudes.

Composition of Air

The mean molar mass of air is 28.97 g/mol. By contrast, *mass-fraction* abundances of gases, particularly for gases with significantly different molecular (molar) mass from that of air will differ from those by volume. For example, in air, helium is 5.2 ppm by *volume-fraction* and *mole-fraction*, but only about (4/29) × 5.2 ppm = 0.72 ppm by *mass-fraction*.

Below the turbopause at an altitude of about 100 km, the Earth's atmosphere has a more-or-less uniform composition as described above; this constitutes the homosphere. However, above about 100 km, the Earth's atmosphere begins to have a composition which varies with altitude. This is essentially because, in the absence of mixing, the density of a gas falls off exponentially with increasing altitude, but at a rate which depends on the molar mass. Thus higher mass constituents, such as oxygen and nitrogen, fall off more quickly than lighter constituents such as helium, molecular hydrogen, and atomic hydrogen. Thus there is a layer, called the heterosphere, in which the earth's atmosphere has varying composition. As the altitude increases, the atmosphere is dominated successively by helium, molecular hydrogen, and atomic hydrogen. The precise altitude of the heterosphere and the layers it contains varies significantly with temperature. After loss of the hydrogen, helium and other hydrogen-containing gases from early Earth due to the Sun's radiation, primitive Earth was devoid of an atmosphere. The first atmosphere was formed by outgassing of gases trapped in the interior of the early Earth, which still goes on today in volcanoes.

Density and Mass

The density of air at sea level is about 1.2 kg/m^3. Natural variations of the barometric pressure occur at any one altitude as a consequence of weather. This variation is relatively small for inhabited altitudes but much more pronounced in the outer atmosphere and space due to variable solar radiation. The atmospheric density decreases as the altitude increases. This variation can be approximately modelled using the barometric formula. More sophisticated models are used by meteorologists and space agencies to predict weather and orbital decay of satellites.

The average mass of the atmosphere is about 5,000 trillion metric tons or 1/1,200,000 the mass of Earth. According to the National Centre for Atmospheric Research, "The total mean mass of the atmosphere is 5.1480×10^{18} kg with an annual range due to water vapour of 1.2 or 1.5×10^{15} kg depending on whether surface pressure or water vapour data are used; somewhat smaller than the previous estimate. The mean mass of water vapour is estimated as 1.27×10^{16} kg and the dry air mass as $5.1352 \pm 0.0003 \times 10^{18}$ kg."

Climate and Weather

The atmosphere of the Earth serves as a key factor in sustaining the planetary ecosystem. The thin layer of gases that envelops the Earth is held in place by the planet's gravity. Dry air consists of 78 per cent nitrogen, 21 per cent oxygen, 1 per cent argon and other inert gases, carbon dioxide, etc.; but air also contains a variable amount of water vapour. The atmospheric pressure declines steadily with altitude, and has a scale height of about 8 kilometres at the Earth's surface: the height at which the atmospheric pressure has declined by a factor of e (a mathematical constant equal to 2.71). The ozone layer of the Earth's atmosphere plays an important role in depleting the amount of ultraviolet (UV) radiation that reaches the surface. As DNA is readily damaged by UV light, this serves to protect life at the surface. The atmosphere also retains heat during the night, thereby reducing the daily temperature extremes.

Terrestrial weather occurs almost exclusively in the lower part of the atmosphere, and serves as a convective system for redistributing heat. Ocean currents are another important factor in determining climate, particularly the major underwater thermohaline circulation which distributes heat energy from the equatorial oceans to the polar regions. These currents help to moderate the differences in temperature between winter and summer in the temperate zones. Also, without the redistributions of heat energy by the ocean currents and atmosphere, the tropics would be much hotter, and the polar regions much colder.

Weather can have both beneficial and harmful effects. Extremes in weather, such as tornadoes or hurricanes and cyclones, can expend large amounts of energy along their paths, and produce devastation. Surface vegetation has evolved a dependence on the seasonal variation of the weather, and sudden changes lasting only a few years can have a dramatic effect, both on the vegetation and on the animals dependent on its growth for their food. The planetary climate is a measure of the long-term trends in the weather. Various factors are known to influence the climate, including ocean currents, surface albedo, greenhouse gases, variations in the solar luminosity, and changes to the planet's orbit. Based on historical records, the Earth is known to have undergone drastic climate changes in the past, including ice ages.

The climate of a region depends on a number of factors, especially latitude. A latitudinal band of the surface with similar climatic attributes forms a climate region. There are a number of such regions, ranging from the tropical climate at the equator to the polar climate in the northern and southern extremes. Weather is also influenced by the seasons, which result from the Earth's axis being tilted relative to its orbital plane. Thus, at any given time during the summer or winter, one part of the planet is more directly exposed to the rays of the sun. This exposure alternates as the Earth revolves in its orbit. At any given time, regardless of season, the northern and southern hemispheres experience opposite seasons.

Weather is a chaotic system that is readily modified by small changes to the environment, so accurate weather forecasting is currently limited to only a few days. Overall, two things are currently happening worldwide: (*a*) temperature is increasing on the average; and (*b*) regional climates have been undergoing noticeable changes.

Hydrosphere

The hydrosphere is the collective mass of water found on, under, and over the surface of a planet. It consists of water in all forms: the ocean, other surface waters including inland seas, lakes, and rivers; rain; underground water; ice; and atmospheric water vapour. The average depth of the oceans is 3,794 m, more than five times the average height of the continents. The mass of the oceans is approximately 1.35×10^{18} tonnes, or about 1/4400 of the total mass of the Earth.

The abundance of water on Earth is a unique feature that distinguishes our "Blue Planet" from others in the solar system. Approximately 70.8 per cent of the Earth is covered by water and only 29.2 per cent is landmass. Earth's solar orbit, volcanism, gravity, greenhouse effect, magnetic field and oxygen-rich atmosphere seem to combine to make Earth a water planet.

Earth is actually beyond the outer edge of the orbits which would be warm enough to form liquid water. Without some form of a greenhouse effect, Earth's water would freeze. Paleontological evidence indicates that at one point after blue-green bacteria had colonised the oceans, the greenhouse effect failed, and Earth's oceans may have completely frozen over for 10 to 100 million years in what is called a snowball Earth event.

On other planets, such as Venus, gaseous water is destroyed by solar ultraviolet radiation, and the hydrogen is ionised and blown away by the solar wind. This effect is slow, but inexorable. This is one hypothesis explaining why Venus has no water. Without hydrogen, the oxygen interacts with the surface and is bound up in solid minerals.

In the Earth's atmosphere, a tenuous layer of ozone within the stratosphere absorbs most of this energetic ultraviolet radiation high in the atmosphere, reducing the cracking effect. The ozone, too, can only be produced in an atmosphere with a large amount of free diatomic oxygen, and so also is dependent on the biosphere. The magnetosphere also shields the ionosphere from direct scouring by the solar wind.

Volcanism continuously emits water vapour from the interior. Earth's plate tectonics recycle carbon and water as limestone rocks are subducted into the mantle and volcanically released as gaseous carbon dioxide and steam. It is estimated that the minerals in the mantle may contain as much as 10 times the water as in all of the current oceans, though most of this trapped water will never be released.

The water cycle describes the methods of transport for water in the hydrosphere. This cycle includes water beneath the Earth's surface and in rocks (lithosphere), the water in plants and animals (biosphere), the water covering the surface of the planet in liquid and solid forms, and the water in the atmosphere in the form of water vapour, clouds, and precipitation. Movement of water within the hydrosphere is described by the hydrologic cycle. It is easy to see this motion in rivers and streams, but it is harder to tell that there is this motion in lakes and ponds.

The water in the oceans moves as it is of different temperature and salinity on different locations. Surface waters are also moved by winds, giving rise to surface ocean currents. Warm water is lighter or less dense than cold water which is more dense or heavier and salty water is also more dense than fresh water. The combination of the water's temperature and salinity determines whether it rises to the surface, sinks to the bottom, or stays at some intermediate depth.

Water Cycle

The Earth's water is always in movement, and the water cycle, also known as the hydrologic cycle, describes the continuous movement of water on, above, and below the surface of the Earth. Since the water cycle is truly a "cycle," there is no beginning or end. Water can change states among liquid, vapour, and ice at various places in the water cycle, with these processes happening in the blink of an eye and over millions of years. Although the balance of water on Earth remains fairly constant over time, individual water molecules can come and go in a hurry.

The water cycle has no starting or ending point. The sun, which drives the water cycle, heats water in the oceans. Some of it evaporates as vapour into the air. Ice and snow can sublimate directly into water vapour. Rising air currents take the vapour up into the atmosphere, along with water from evapotranspiration, which is water transpired from plants and evaporated from the soil. The vapour rises into the air where cooler temperatures cause it to condense into clouds. Air currents move clouds around the globe, cloud particles collide, grow, and fall out of the sky as precipitation. Some precipitation falls as snow and can accumulate as ice caps and glaciers, which can store frozen water for thousands of years. Snowpacks in warmer climates often thaw and melt when spring arrives, and the melted water flows overland as snowmelt.

Figure 1.1. Water Cylce

Most precipitation falls back into the oceans or onto land, where, due to gravity, the precipitation flows over the ground as surface runoff. A portion of runoff enters rivers in valleys in the landscape, with streamflow moving water towards the oceans. Runoff, and groundwater seepage, accumulate and are stored as freshwater in lakes. Not all runoff flows into rivers. Much of it soaks into the ground as infiltration. Some water infiltrates deep into the ground and replenishes aquifers (saturated subsurface rock), which store huge amounts of freshwater for long periods of time. Some infiltration stays close to the land surface and can seep back into surface-water bodies (and the ocean) as groundwater discharge, and some ground water finds openings in the land surface and emerges as freshwater springs. Over time, the water continues flowing, some to reenter the ocean, where the water cycle renews itself.

The water cycle describes the processes that drive the movement of water throughout the hydrosphere. However, much more water is "in storage" for long periods of time than is actually moving through the cycle. The storehouses for the vast majority of all water on Earth are the oceans. It is estimated that of the 332,500,000 cubic miles of the world's water supply, about 321,000,000 mi^3 is stored in oceans, or about 95 per cent. It is also estimated that the oceans supply about 90 per cent of the evaporated water that goes into the water cycle.

During colder climatic periods more ice caps and glaciers form, and enough of the global water supply accumulates as ice to lessen the amounts in other parts of the water cycle. The reverse is true during warm periods. During the last ice age glaciers covered almost one-third of Earth's land mass, with the result being that the oceans were about 122 metres lower than today. During the last global "warm spell," about 125,000 years ago, the seas were about 5.5. metres higher than they are now. About three million years ago the oceans could have been up to 50 metres higher.

The scientific consensus expressed in the 2007 Intergovernmental Panel on Climate Change (IPCC)

Summary for Policymakers is for the water cycle to continue to intensify throughout the 21st century, though this does not mean that precipitation will increase in all regions. In subtropical land areas precipitation is projected to decrease during the 21st century, increasing the probability of drought. The drying is projected to be strongest near the poleward margins of the subtropics. Annual precipitation amounts are expected to increase in near-equatorial regions that tend to be wet in the present climate, and also at high latitudes. These large-scale patterns are present in nearly all of the climate model simulations conducted at several international research centres as part of the 4th Assessment of the IPCC.

Glacial retreat is also an example of a changing water cycle, where the supply of water to glaciers from precipitation cannot keep up with the loss of water from melting and sublimation. Glacial retreat since 1850 has been extensive.

Effects on Climate

The water cycle is powered from solar energy. 86 per cent of the global evaporation occurs from the oceans, reducing their temperature by evaporative cooling. Without the cooling effect of evaporation the greenhouse effect would lead to a much higher surface temperature of 67°C, and a warmer planet.

Most of the solar energy warms tropical seas. After evaporating, water vapour rises into the atmosphere and is carried by winds away from the tropics. Most of this vapour condenses as rain in the Intertropical convergence zone (ITCZ), releasing latent heat that warms the air. This in turn drives the atmospheric circulation.

While the water cycle is itself a biogeochemical cycle, flow of water over and beneath the Earth is a key component of the cycling of other biogeochemicals. Runoff is responsible for almost all of the transport of eroded sediment and phosphorus from land to waterbodies. The salinity of the oceans is derived from erosion and transport of dissolved salts from the land. Cultural eutrophication of lakes is primarily due to phosphorus, applied in excess to agricultural fields in fertilisers, and then transported overland and down rivers. Both runoff and groundwater flow play significant roles in transporting nitrogen from the land to waterbodies.

Lithosphere

The lithosphere is the solid outermost shell of a rocky planet. On the Earth, the lithosphere includes the crust and the uppermost mantle which is joined to the crust across the Mohorovicic discontinuity. Lithosphere is underlain by asthenosphere, the weaker, hotter, and deeper part of the upper mantle. The base of the lithosphere-asthenosphere boundary corresponds approximately to the depth of the melting temperature in the mantle. As the conductively cooling surface layer of the Earth's convection system, the lithosphere thickens over time. It is fragmented into tectonic plates, which move independently relative to one another. This movement of lithospheric plates is described as plate tectonics.

The concept of the lithosphere as Earth's strong outer layer was developed by Barrell, who wrote a series of papers introducing the concept. The concept was based on the presence of significant gravity anomalies over continental crust, from which he inferred that there must exist a strong upper layer above a weaker layer which could flow. These ideas were enlarged by Daly, and have been broadly accepted by geologists and geophysicists.

Although these ideas about lithosphere and asthenosphere were developed long before plate tectonic theory was articulated in the 1960's, the concepts that strong lithosphere exists and that this rests on weak asthenosphere are essential to that theory.

The division of Earth's outer layers into lithosphere and asthenosphere should not be confused with the chemical subdivision of the outer Earth into mantle, and crust. All crust is in the lithosphere, but lithosphere generally contains more mantle than crust.

There are two types of lithosphere:

1. Oceanic lithosphere, which is associated with oceanic crust.
2. Continental lithosphere, which is associated with Continental crust.

Oceanic lithosphere is typically about 50-100 km thick, while continental lithosphere is about 150 km thick, consisting ~50 km of crust and 100 km or more of uppermost mantle. Oceanic lithosphere consists mainly of mafic crust and ultramafic mantle and is denser than continental lithosphere, for which the mantle is associated with crust made of felsic rocks. The crust is distinguished from the upper mantle by the change in chemical composition that takes place at the Moho discontinuity. Oceanic lithosphere thickens as it ages and moves away from the mid-ocean ridge. This thickening occurs by conductive cooling, which converts hot asthenosphere into lithospheric mantle, and causes the oceanic lithosphere to become increasingly dense with age. Oceanic lithosphere is less dense than asthenosphere for a few tens of millions of years, but after this becomes increasingly denser than asthenosphere. The gravitational instability of mature oceanic lithosphere has the effect that at subduction zones the oceanic lithosphere invariably sinks underneath the overriding lithosphere, which can be oceanic or continental. New oceanic lithosphere is constantly being produced at mid-ocean ridges and is recycled back to the mantle at subduction zones. As a result, oceanic lithosphere is much younger than continental lithosphere: the oldest oceanic lithosphere is about 170 million years old, while parts of the continental lithosphere are billions of years old.

Another distinguishing characteristic of the lithosphere is its flow properties. Under the influence of the low-intensity, long-term stresses that drive plate tectonic motions, the lithosphere responds essentially as a rigid shell and thus deforms primarily through brittle failure, whereas the asthenosphere is heat-softened and accommodates strain through plastic deformation.

Rock

Rock is a naturally occurring aggregate of minerals and/ or mineraloids. The Earth's lithosphere is made of rock. In general rocks are of three types, namely, igneous, sedimentary, and metamorphic. Petrology is the scientific study of rocks. Rocks are classified by mineral and chemical composition, by the texture of the constituent particles and by the processes that formed them. These indicators separate rocks into igneous, sedimentary and metamorphic. They may also be classified according to particle size, in the case of conglomerates and breccias or in the case of individual **stones**. The transformation of one rock type to another is described by the geological model called the rock cycle.

Igneous rocks are formed when molten magma cools and are divided into two main categories: plutonic rock and volcanic. Plutonic or intrusive rocks result when magma cools and crystallises slowly within the Earth's crust, while volcanic or extrusive rocks result from magma reaching the surface either as lava or *fragmental ejecta*.

Sedimentary rocks are formed by deposition of either clastic sediments, organic matter, or chemical precipitates, followed by compaction of the particulate matter and cementation during diagenesis. Sedimentary rocks form at or near the Earth's surface. Mud rocks comprise 65 per cent; sandstones 20 to 25 per cent and carbonate rocks 10 to 15 per cent.

Metamorphic rocks are formed by subjecting any rock type to different temperature and pressure conditions than those in which the original rock was formed. These temperatures and pressures are always higher than those at the Earth's surface and must be sufficiently high so as to change the original minerals into other mineral types or else into other forms of the same minerals. There are also Foliated and nonfoliated rock types.

The three classes of rocks: the igneous, the sedimentary and the metamorphic — are subdivided into many groups. There are, however, no hard and fast boundaries between allied rocks. By increase or diminution in the proportions of their constituent minerals they pass by every gradation into one another, the distinctive structures also of one kind of rock may often be traced gradually merging into those of another. Hence the definitions adopted in establishing rock nomenclature merely correspond to selected points in a continuously graduated series. This is frequently urged as a reason for reducing rock classification to its simplest possible terms, and using only a few generalised rock designations. But it is clear that many apparently trivial differences tend regularly to recur, and have a real significance, and so long as any variation can be shown to be of this nature it deserves recognition.

Iron oxides and carbonates play a large part in many sedimentary rocks and are especially important as colouring agents. The red sands and limestones, for example, which are so abundant, contain small amounts of iron (III) oxide (hematite), which in a finely divided state gives a red hue to all rocks in which it is present. Limonite and goethite, on the other hand, makes rocks yellow or brown; manganese oxides, asphalt and other carbonaceous substances are the cause of the black colour of many sediments.

Soil

Soil is a naturally occurring, unconsolidated or loose material on the surface of the earth, capable of supporting life. It has three components: solid, liquid, and gas. The solid phase is a mixture of mineral and organic matter. Soil particles pack loosely, forming a soil structure filled with voids. The solid phase occupies about half of the soil volume. The remaining void space contains water (liquid) and air (gas).

Soil colour is the first impression one has when viewing soil. Striking colours and contrasting patterns are especially memorable. The Red River in Mississippi carries sediment eroded from extensive reddish soils like Port Silt Loam in Oklahoma. The Yellow River in China carries yellow sediment from eroding loessal soils. Mollisols in the Great Plains are darkened and enriched by organic matter. Podsols in boreal forests have highly contrasting layers due to acidity and leaching.

Soil colour results from chemical weathering. As the primary minerals in parent material weather, the elements combine into new and colourful compounds. Iron forms secondary minerals with a yellow or red colour; organic matter decomposes into black and brown compounds; and manganese forms black mineral deposits.

Soil structure is the arrangement of soil particles into aggregates. These may have various shapes, sizes and degrees of development or expression.

Soil texture refers to sand, silt and clay composition. Sand and silt are the product of physical weathering while clay is the product of chemical weathering. Clay content is particularly influential on soil behaviour due to a high retention capacity for nutrients and water.

Soil Formation

Soil formation, or pedogenesis, is the combined effect of physical, chemical, biological, and anthropogenic processes on soil parent material resulting in the formation of soil horizons. Soil is always changing. The long periods over which change occurs and the multiple influences of change mean that simple soils are rare. While soil can achieve relative stability in properties for extended periods of time, the soil life cycle ultimately ends in soil conditions that leave it vulnerable to erosion. Little of the soil continuum of the earth is older than Tertiary and most no older than Pleistocene. Despite the inevitability of soils retrogression and degradation, most soil cycles are long and productive. How the soil "life" cycle proceeds is influenced by at least five classic soil forming factors: regional climate, biotic potential, topography, parent material, and the passage of time.

Geologists have a particular interest in the patterns of soil on the surface of the earth. Soil texture, colour and chemistry often reflect the underlying geologic parent material and soil types often change at geologic unit boundaries. Buried paleosols mark previous land surfaces and record climatic conditions from previous eras. Geologists use this paleopedological record to understand the ecological relationships in past ecosystems. According to the theory of biorhexistasy, prolonged conditions conducive to forming deep, weathered soils result in increasing ocean salinity and the formation of limestone.

Geologists use soil profile features to establish the duration of surface stability in the context of geologic faults or slope stability. An offset subsoil horizon indicates rupture during soil formation and the degree of subsequent subsoil formation is relied upon to establish time since rupture. Soil examined in shovel test pits is used by archaeologists for relative dating based on stratigraphy. What is considered most typical is to use soil profile features to determine the maximum reasonable pit depth than needs to be examined for archaeological evidence in the interest of cultural resources management.

Uses

Soil material is a critical component in the mining and construction industries. Soil serves as a foundation for most construction projects. Massive volumes of soil can be involved in surface mining, road building, and dam construction. Earth sheltering is the architectural practice of using soil for external thermal mass against building walls. Soil resources are critical to the environment, as well as to food and fibre production. Waste management often has a soil component. For example, septic drain fields treat septic tank effluent uses aerobic soil processes. Organic soils, especially peat, serve as a significant fuel resource.

Biosphere

The biosphere is the part of the Earth, including air, land, surface rocks, and water, within which life occurs, and which biotic processes in turn alter or transform. From the broadest biophysiological point of view, the biosphere is the global ecological system integrating all living beings and their relationships, including their interaction with the elements of the lithosphere, hydrosphere, and atmosphere. This biosphere is postulated to have evolved, beginning through a process of biogenesis or biopoesis, at least some 3.5 billion years ago.

Biomass accounts for about 3.7 kg carbon per square metre of the earth's surface averaged over land and sea, making a total of about 1900 gigatonnes of carbon. Nearly every part of the planet, from the polar ice caps to the Equator, supports life of some kind. Recent advances in microbiology have demonstrated that microbes live deep beneath the Earth's terrestrial surface, and that the total mass of microbial life in so-called "uninhabitable zones" may, in biomass, exceed all animal and plant life on the surface. The actual thickness of the biosphere on earth is hard to measure. Birds typically fly at altitudes of 650 to 2000 metres, and fish that live deep underwater can be found down to -8,372 metres in the Puerto Rico Trench.

Microscopic organisms live at such extremes that, taking them into consideration puts the thickness of the biosphere much greater, but at minimum it extends from 5,400 metres above sea level to at least 9,000 metres below sea level.

When the word *Biosphere* is followed by a number, it is usually referring to a specific system. Thus:

1. *Biosphere 1*: The planet Earth.
2. *Biosphere 2*: A laboratory in Arizona which contains 3.15 acres (13,000 m^2) of closed ecosystem.
3. *Biosphere 3*: Experiment to be conducted starting Jan 2008 in the Gary C. Comer Geochemistry building at Columbia University's Lamont Campus in Palisades, New York.
4. *Biosphere J*: An experiment in Japan.

Our biosphere is divided into a number of biomes, inhabited by broadly similar flora and fauna. On land, biomes are separated primarily by latitude. Terrestrial biomes lying within the Arctic and Antarctic Circles are relatively barren of plant and animal life, while most of the more populous biomes lie near the Equator. Terrestrial organisms in temperate and Arctic biomes have relatively small amounts of total biomass, smaller energy budgets, and display prominent adaptations to cold, including world-spanning migrations, social adaptations, homeothermy, estivation and multiple layers of insulation.

Life

Although there is no universal agreement on the definition of life, scientists generally accept that the biological manifestation of life is characterised by organisation, metabolism, growth, adaptation, response to stimuli and reproduction. Life may also be said to be simply the characteristic state of organisms.

Properties common to terrestrial organisms are that they are cellular, carbon-and-water-based with complex organisation, having a metabolism, a capacity to grow, respond to stimuli, and reproduce. An entity with these properties is generally considered life. However, not every definition of life considers all of these properties to be essential. Human-made analogs of life may also be considered to be life.

The biosphere is the part of Earth's outer shell—including air, land, surface rocks and water—within which life occurs, and which biotic processes in turn alter or transform. From the broadest geophysiological point of view, the biosphere is the global ecological system integrating all living beings and their relationships, including their interaction with the elements of the lithosphere (rocks), hydrosphere (water), and atmosphere (air). Currently the entire Earth contains over 75 billion tons of biomass (life), which lives within various environments within the biosphere.

Over nine-tenths of the total biomass on Earth is plant life, on which animal life depends very heavily for its existence. More than 2 million species of plant and animal life have been identified to date, and estimates of the actual number of existing species range from several million to well over 50 million. The number of individual species of life is constantly in some degree of flux, with new species appearing and others ceasing to exist on a continual basis. The total number of species is presently in rapid decline.

Life, as we understand it, is currently only known to exist on the planet Earth. The origin of life is still a poorly understood process, but it is thought to have occurred about 3.9 to 3.5 billion years ago during the hadean or archean eons on a primordial earth that had a substantially different environment than is found at present. These life forms possessed the basic traits of self-replication and inheritable traits. Once life had appeared, the process of evolution by natural selection resulted in the formation of ever-more diverse life forms.

Species that were unable to adapt to the changing environment and competition from other life forms became extinct. However, the fossil record retains evidence of many of these older species. Current fossil and DNA evidence shows that all existing species can trace a continual ancestry back to the first primitive life forms.

The advent of photosynthesis in very basic forms of plant life worldwide allowed the sun's energy to be harvested to create conditions allowing for more complex life. The resultant oxygen accumulated in the atmosphere and gave rise to the ozone layer. The incorporation of smaller cells within larger ones resulted in the development of yet more complex cells called eukaryotes. Cells within colonies became increasingly specialised, resulting in true multicellular organisms. With the ozone layer absorbing harmful ultraviolet radiation, life colonised the surface of Earth.

Microbes

The first form of life to develop on the Earth were microbes, and they remained the only form of life on the planet until about a billion years ago when multi-cellular organisms began to appear. Microorganisms are single-celled organisms that are generally smaller than the human eye can see. They include Bacteria, Fungi, Archaea and Protista.

These life forms are found in almost every location on the Earth where there is liquid water, including the interior of rocks within the planet. Their reproduction is both rapid and profuse. The combination of a high mutation rate and a horizontal gene transfer ability makes them highly adaptable, and able to survive in new environments, including outer space. They form an essential part of the planetary ecosystem. However some microorganisms are pathogenic and can post health risk to other organisms.

Plants and Animals

The distinction between plant and animal life is not sharply drawn, with some categories of life that stand between or across the two. Originally Aristotle divided all living things between plants, which generally do not move, and animals. In Linnaeus' system, these became the Kingdoms Vegetabilia and Animalia. Since then, it has become clear that the Plantae as originally defined included several unrelated groups, and the fungi and several groups of algae were removed to new kingdoms. However, these are still often considered plants in many contexts. Bacterial life is sometimes included in flora, and some classifications use the term *bacterial flora* separately from *plant flora.*

Among the many ways of classifying plants are by regional floras, which, depending on the purpose of study, can also include *fossil flora,* remnants of plant life from a previous era. People in many regions and countries take great pride in their individual arrays of characteristic flora, which can vary widely across the globe due to differences in climate and terrain.

Regional floras commonly are divided into categories such as *native flora* and *agricultural and garden flora,* the latter of which are intentionally grown and cultivated. Some types of "native flora" actually have been introduced centuries ago by people migrating from one region or continent to another, and become an integral part of the native, or natural flora of the place to which they were introduced. This is an example of how human interaction with nature can blur the boundary of what is considered nature.

Another category of plant has historically been carved out for *weeds*. Though the term has fallen into disfavour among botanists as a formal way to categorise "unclean" plants, the informal use of the word "weeds" to describe those plants that are deemed worthy of elimination is illustrative of the general tendency of people and societies to seek to alter or shape the course of nature. Similarly, animals are often categorised in ways such as *domestic, farm animals, wild animals, pests,* etc. according to their relationship to human life.

Animals as a category have several characteristics that generally set them apart from other living things, though not traced by scientists to having legs or wings instead of roots and leaves. Animals are eukaryotic and usually multicellular, which separates them from bacteria, archaea and most protists. They are heterotrophic, generally digesting food in an internal chamber, which separates them from plants and algae. They are also distinguished from plants, algae, and fungi by lacking cell walls.

With a few exceptions, most notably the sponges, animals have bodies differentiated into separate tissues. These include muscles, which are able to contract and control locomotion, and a nervous system, which sends and processes signals. There is also typically an internal digestive chamber. The eukaryotic cells possessed by all animals are surrounded by a characteristic extracellular matrix composed of collagen and elastic glycoproteins. This may be calcified to form structures like shells, bones, and spicules, a framework upon which cells can move about and be reorganised during development and maturation, and which supports the complex anatomy required for mobility.

Ecosystems

All forms of life interact with the environment in which they exist, and also with other life forms. In the 20th century this premise gave rise to the concept of **ecosystems**, which can be defined as any situation where there is interaction between organisms and their environment. Ecosystems are composed of a variety of abiotic and biotic components that function in an interrelated way. The structure and composition is determined by various environmental factors that are interrelated. Variations of these factors will initiate dynamic modifications to the ecosystem. Some of the more important components are: soil, atmosphere, radiation from the sun, water, and living organisms.

Each living organism has a continual relationship with every other element that makes up its environment. Within the ecosystem, species are connected and dependent upon one another in the food chain, and exchange energy and matter between themselves as well as with their environment. Every species has limits of tolerance to factors that affect its survival, reproductive success and ability to continue to thrive and interact sustainably with the rest of its environment, which in turn may have effects on these factors for many other species or even on the whole of life. The concept of an ecosystem is thus an important subject of study, as such study provides information needed to make decisions about how human life may interact in a way that allows the various ecosystems to be sustained for future use rather than used up or otherwise rendered ineffective. For the purpose of such study, a unit of smaller size is called a microecosystem. For example, an ecosystem can be a stone and all the life under it. A *macroecosystem* might involve a whole ecoregion, with its drainage basin.

MAN-ENVIRONMENTAL RELATIONSHIP

Although humans currently comprise only about one-half of one percent of the total living biomass on Earth, the human effect on nature is disproportionately large. Because of the extent of human influence, the boundaries between what we regard as nature and "made environments" is not clear cut except at the extremes. Even at the extremes, the amount of natural environment that is free of discernible human influence is presently diminishing at an increasingly rapid pace, or, according to some, has already disappeared. The development of technology by the human race has allowed the greater exploitation of natural resources and has helped to alleviate some of the risk from natural hazards. In spite of this progress, however, the fate of human civilisation remains closely linked to changes in the environment. There exists a highly complex feedback-loop between the use of advanced technology and changes to the environment that are only slowly becoming understood. Manmade threats to the Earth's natural environment include pollution, deforestation, and disasters such as oil spills. Humans have contributed to the extinction of many plants and animals.

Humans employ nature for both leisure and economic activities. The acquisition of natural resources for industrial use remains a primary component of the world's economic system. Some activities, such as hunting and fishing, are used for both sustenance and leisure, often by different people. Agriculture was first adopted around the 9th millennium BCE. Ranging from food production to energy, nature influences economic wealth.

Although early humans gathered uncultivated plant materials for food and employed the medicinal properties of vegetation for healing, most modern human use of plants is through agriculture. The clearance of large tracts of land for crop growth has led to a significant reduction in the amount available of forestation and wetlands, resulting in the loss of habitat for many plant and animal species as well as increased erosion.

2

THE ECOSYSTEM

The study of ecosystems mainly consists of the study of certain processes that link the living, or biotic, components to the non-living, or abiotic, components. *Energy transformations* and *biogeochemical cycling* are the main processes that comprise the field of ecosystem ecology. Ecology generally is defined as the interactions of organisms with one another and with the environment in which they occur. We can study ecology at the level of the individual, the population, the community, and the ecosystem.

Studies of *individuals* are concerned mostly about physiology, reproduction, development or behaviour, and studies of *populations* usually focus on the habitat and resource needs of individual species, their group behaviours, population growth, and what limits their abundance or causes extinction. Studies of *communities* examine how populations of many species interact with one another, such as predators and their prey, or competitors that share common needs or resources.

In *ecosystem ecology* we put all of this together and, insofar as we can, we try to understand how the system operates as a whole. This means that, rather than worrying mainly about particular species, we try to focus on major functional aspects of the system. These *functional aspects* include such things as the amount of energy that is produced by photosynthesis, how energy or materials flow along the many steps in a food chain, or what controls the rate of decomposition of materials or the rate at which nutrients are recycled in the system.

PARTS OF AN ECOSYSTEM

We can clarify the parts of an ecosystem by listing them under the headings "abiotic" and "biotic".

Abiotic Components	*Biotic Components*
Sunlight	Primary producers
Temperature	Herbivores
Precipitation	Carnivores
Water or moisture	Omnivores
Soil or water chemistry (e.g., P, NH_4^+) etc.	Detritivores etc.
All of these vary over space/time	

By and large, this set of environmental factors is important almost everywhere, in all ecosystems. Usually, biological communities include the "functional groupings". A *functional group* is a biological category composed of organisms that perform mostly the same kind of function in the system; for example, all the photosynthetic plants or primary producers form a functional group. Membership in the functional group does not depend very much on who the actual players happen to be, only on what function they perform in the ecosystem.

Functions of Ecosystems

This figure with the plants, zebra, lion, and so forth illustrates the two main ideas about how ecosystems function: *ecosystems have energy flows* and *ecosystems cycle materials*. These two processes are linked, but they are not quite the same (Figure 2.1).

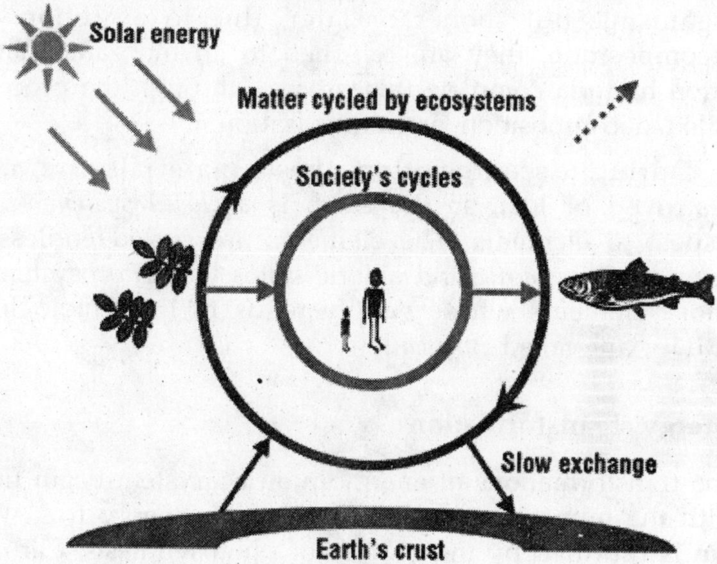

Figure 2.1. Energy Flow

Energy enters the biological system as light energy, or photons, is transformed into chemical energy in organic molecules by cellular processes including photosynthesis and respiration, and ultimately is converted to heat energy. This energy is dissipated, meaning it is lost to the system as heat; once it is lost it cannot be recycled. Without the continued input of solar energy, biological systems would quickly shut down. Thus the earth is an *open system* with respect to energy. Elements such as carbon, nitrogen, or phosphorus enter living organisms in a variety of ways.

Plants obtain elements from the surrounding atmosphere, water, or soils. Animals may also obtain elements directly from the physical environment, but usually they obtain these mainly as a consequence of consuming other organisms. These materials are transformed biochemically within the bodies of organisms, but sooner or later, due to excretion or decomposition, they are returned to an inorganic state. Often bacteria complete this process, through the process called decomposition or mineralisation.

During decomposition these materials are not destroyed or lost, so the earth is a *closed system* with respect to elements. The elements are cycled endlessly between their biotic and abiotic states within ecosystems. Those elements whose supply tends to limit biological activity are called *nutrients*.

Energy Transformation

The transformations of energy in an ecosystem begin first with the input of energy from the sun. Energy from the sun is captured by the process of photosynthesis. Carbon dioxide is combined with hydrogen to produce carbohydrates (CHO). Energy is stored in the high energy bonds of adenosine triphosphate, or ATP.

The prophet Isaah said "all flesh is grass", earning him the title of first ecologist, because virtually all energy available to organisms originates in plants. Because it is the first step in the production of energy for living things, it is called *primary production*. *Herbivores* obtain their energy by consuming plants or plant products, *carnivores* eat herbivores, and *detritivores* consume the droppings and carcasses of us all.

Figure 2.2 portrays a food chain, in which energy from the sun, captured by plant photosynthesis, flows from *trophic level* to trophic level via the *food chain*. A trophic level is composed of organisms that make a living in the same way, that is they are all *primary producers* (plants), *primary consumers* (herbivores) or *secondary consumers* (carnivores).

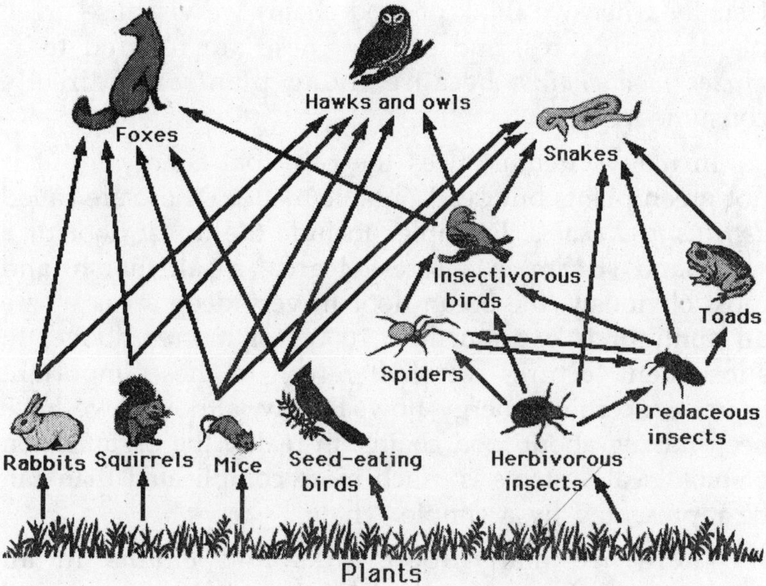

Figure 2.2.

Dead tissue and waste products are produced at all levels. Scavengers, detritivores, and decomposers collectively account for the use of all such "waste" — consumers of carcasses and fallen leaves may be other animals, such as crows and beetles, but ultimately it is the microbes that finish the job of decomposition. Not surprisingly, the amount of primary production varies a great deal from place to place, due to differences in the amount of solar radiation and the availability of nutrients and water.

For reasons that we will explore more fully in subsequent lectures, *energy transfer through the food chain is inefficient*. This means that less energy is available at the herbivore level than at the primary producer level, less yet at the carnivore level, and so on. The result is a pyramid of energy, with important implications for understanding the quantity of life that can be supported.

Usually when we think of food chains we visualise green plants, herbivores, and so on. These are referred to as *grazer food chains*, because living plants are directly consumed.

In many circumstances the principal energy input is not green plants but dead organic matter. These are called *detritus food chains*. Examples include the forest floor or a woodland stream in a forested area, a salt marsh, and most obviously, the ocean floor in very deep areas where all sunlight is extinguished 1000's of metres above. In subsequent lectures we shall return to these important issues concerning energy flow. Finally, although we have been talking about food chains, in reality the organisation of biological systems is much more complicated than can be represented by a simple "chain".

There are many food links and chains in an ecosystem, and we refer to all of these linkages as a *food web*. Food webs can be very complicated, where it appears that *"everything is connected to everything else"*, and it is important to understand what are the most important linkages in any particular food web.

Control of Ecosystem

There are two dominant theories of the control of ecosystems. The first, called *bottom-up control*, states that it is the nutrient supply to the primary producers that ultimately controls how ecosystems function. If the nutrient supply is increased, the resulting increase in production of autotrophs is propagated through the food web and all of the other trophic levels will respond to the increased availability of food.

The second theory, called *top-down control*, states that predation and grazing by higher trophic levels on lower trophic levels ultimately controls ecosystem function. For example, if you have an increase in predators, that increase will result in fewer grazers, and that decrease in grazers will result in turn in more primary producers because fewer of them are being eaten by the grazers. Thus the control of population numbers and overall productivity "cascades" from the top levels of the food chain down to the bottom trophic levels.

There is evidence from many ecosystem studies that both controls are operating to some degree, but that neither control is complete. For example, the "top-down" effect is often very strong at trophic levels near to the top predators, but the control weakens as you move further down the food chain.

Similarly, the "bottom-up" effect of adding nutrients usually stimulates primary production, but the stimulation of secondary production further up the food chain is less strong or is absent. Thus we find that both of these controls are operating in any system at any time, and we must understand the relative importance of each control in order to help us to predict how an ecosystem will behave or change under different circumstances, such as in the face of a changing climate.

Geography of Ecosystems

There are many different ecosystems: rain forests and tundra, coral reefs and ponds, grasslands and deserts. Climate differences from place to place largely determine the types of ecosystems we see. How terrestrial ecosystems appear to us is influenced mainly by the dominant vegetation

The word "biome" is used to describe a major vegetation type such as tropical rain forest, grassland, tundra, etc., extending over a large geographic area. It is never used for aquatic systems, such as ponds or coral reefs. It always refers to a vegetation category that is dominant over a very large geographic scale, and so is somewhat broader than an ecosystem.

Ecological (Trophic) Pyramid

Ecological Pyramid (or Trophic pyramid) is a graphical representation designed to show the biomass or productivity at each trophic level in a given ecosystem. Biomass pyramids show the abundance or biomass of organisms at each trophic level, while productivity pyramids show the production or turn-over in biomass. Ecological Pyramids begin with producers on the bottom and proceed through the various trophic levels, the highest of which is on top.

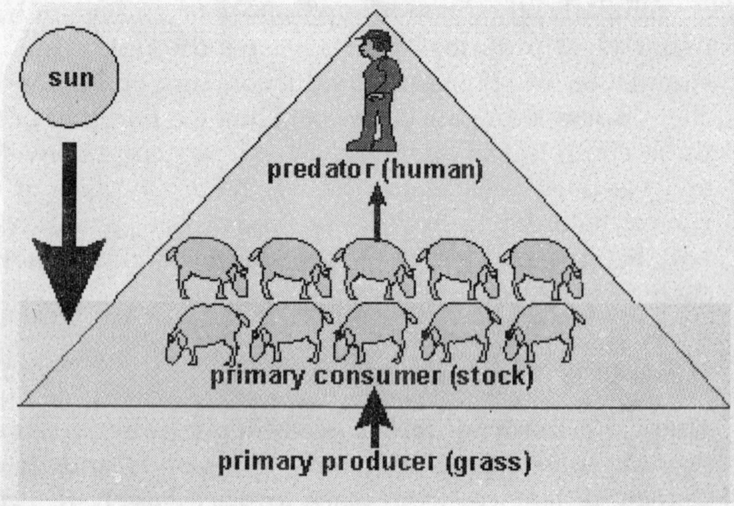

Figure 2.3. Ecological Pyramid

Biomass Pyramid

An Ecological Pyramid of Biomass does shows the relationship between biomass and trophic level by quantifying the amount of biomass present at each trophic level. Typical units for a biomass pyramid could be grams per metre^2, or calories per metre^2. Biomass pyramids provide a single snapshot in time of an ecological community.

One problem with biomass pyramids is that they can make a trophic level look like it contains more energy than it actually does. For example, all birds have a beak and skeleton, which despite taking up mass are not eaten by the next trophic level. In a Pyramid of Biomass, the skeleton and beak would still be quantified even though it does not contribute to the overall flow of energy into the next trophic level.

Biomass pyramids can also be misleading, in that they do not include the turn-over in biomass at a given trophic level. This can lead to inverted pyramids, in which small plakton are able to hunt down andhieat whales.

ECOSYSTEM DIVERSITY

Ecosystem diversity denotes the diversity of habitats, communities and ecological processes within the biosphere. Ecosystem is a natural functional unit. Ecosystem is also a dynamic function unit. When we talk about ecosystem diversity we generally consider its two levels, that is, communities and ecosystems. Considering the global changes of ecosystem and the conservation of endangered species, it is very important to study ecosystem diversity.

Biological Community and Ecosystem

A biological community is defined by the species that occupy a particular locality and interaction between those species. A biological community together with its associated physical environment is termed an ecosystem. Within an ecosystem, water evaporates from biological communities and Earth's surface, to all again as rain or snow and replenish terrestrial and aquatic environment. Soil is built up out of parent rock material and decaying organic matter.

Photosynthetic plants absorb light energy, and this energy is released back into the atmosphere as heat during the lives of plants and animals, as well as after they die and decompose. Plants absorb carbon dioxide and release oxygen during respiration, while animals and fungi absorb oxygen and release carbon dioxide. Mineral nutrients, such as nitrogen and phosphorus, cycle between the living and the non-living compartments of the ecosystem. The physical environment affects the structure and characteristics of a biological community; but the biological community can also have effects on the physical characteristics of the ecosystem.

In a terrestrial ecosystem, for example, wind speed, humidity, and temperature in a given location can all be affected by the plants and animals present there. In aquatic ecosystem, such characteristics as water turbulence and clarity, water chemistry, and water depth affect the characteristics of the biological communities, but in turn biological communities such as kelp forests and coral reefs can affect the physical environment. Within a biological community each species utilises a unique set of resources that constitute its niche.

The niche for a plant species might consist of the type of soil on which it is found, the amount of sunlight it receives, the amount of soil moisture it requires, the type of pollination system it has, and its mechanism of seed dispersal. The niche for an animal might include the type of habitat it occupies, the types of food it eats over the year, and its requirements for overwintering dens. The niche of each species is the space where that species exists in nature and is not outcompeted by other species.

Ecosystem Fluctuation

The niche of a species often includes the stage in succession that the species occupies; succession is the gradual process of change in species composition, community structure, and physical characteristics that occur following natural and human-caused disturbance to a biological community. Particular species are often found at particular successional stages. For example, sun-living butterflies and annual plants are found most commonly early in succession, in the months immediately after a hurricane or a logging operation has destroyed an old-growth forest.

At this stage, the ground is receiving high levels of sunlight, with high temperatures and low humidity during the day. Over the course of years and decades, the forest is gradually reestablished. Different species, including shade-tolerant wildflowers and birds that nest in holes in dead trees, are found in these late-successional stages of mature trees and old-growth forest. Similar patterns of succession-stage specific organisms are found in other biological communities, such as grassland, lakeshores, and the intertidal zones of oceans.

The composition of communities is often strongly affected by competition and predation. Predators often dramatically reduce the numbers of populations of prey species and may eliminate some prey species from certain habitats. Predators may indirectly increase biological diversity in a community by keeping the densities of other species so low that competition for resources does not occur. A good example of this is the marine intertidal ecosystem in which a large sea star (starfish), pisaster, feeds on 15 species of mollusks that cling to the rocks.

As long as the predatory sea star is present, the mollusks species do not compete with each other for space on the rocks, since they are eaten too fast to allow them to achieve high population densities. Under these circumstances, all 15 species are able to occupy the niche. If the sea star is removed, however, the mollusks increase in abundance and start competing for space on the rocks. In the absence of predation, competition between species reduces the numbers of species; eventually only a few of the original 15 species remains, with some rocks taken over by just one species.

In plants communities as well, species diversity is often higher when intense grazing by animals prevents competition among plants than when grazers are absent. The number of individuals of a particular species that the resource of an environment can support is termed the carrying capacity. A population's numbers are often well below the carrying capacity when it is held in check by a predator population. If the predator is removed, the population may increase to a point where it reaches the carrying capacity, or may even increase beyond the carrying capacity to a point at which the environment is damaged and the population crashes.

The population size of a species may also be controlled by other species that compete with it for the same resources; for example, two heron species may eat the same type of fish. Population size may decline if a competing species becomes abundant, or may increase if the competitor becomes scarce or is also eliminated from the community.

Measuring Diversity

Ecologists have measured diversity either by estimating species richness in an area, or by one or more indexes combining species richness and relative abundance within an area. Some attempts have also been made to measure change in species richness between areas. These solutions to the problem of measuring biodiversity are limited because species richness takes no account of the differences between species in relation to their place in the natural hierarchy, and because relative abundance is not a fixed property of species, varying widely from time to time and place to place. Furthermore, in many environments most taxa are virtually or even completely unknown.

In a seminal work on the measurement of diversity, Whittaker introduced the concepts of alpha, beta and gamma diversity. The measurements just described, giving diversity values for single sites, are examples of alpha diversity. The beta and gamma diversity concepts relate to changes in diversity between sites at local (beta) and geographical (gamma) scales. An essential part of these relational concepts is the idea of species turnover - the degree to which species present at one site are replaced by others at different sites. For use in assessing the relative value of multiple sites for the conservation of biodiversity, the idea of species turnover is translated into the principle of complementarity, implemented in combination with a taxonomic diversity index.

Alpha, Beta and Gamma Diversity

There are many formulas to determine alpha or beta diversity of certain community. Generally, the following formulas are used.

Alpha diversity

Numerical species richness: dGI = S/ln A, dMa= (S-1)/ln A and dMo = S/N. In formulas, S is the total numbers of species in study area; A is the area of the study area; N is the total numbers of animals of all species.

Simpson index: In formula, Pi = Ni/N, Ni is the numbers of species i; N is the total numbers of all species.

Shannon-Wiener index: In formula, Pi is the same as in the formula of Simpson index.

Beta diversity

There are three indexes of beta diversity, Whittaker index:, Cody index: $\beta c = [g(H)+l(h)]/2$ and Wilson and Shmida index $\beta T = [g(H)+l(H)]/2a$. In formulas, S is the total numbers of species in study ecosystem; ma is the average numbers of species in individual study area;g (H) is the increasing numbers of species along with changes of habitat grads; l(H) is the decreasing numbers of species along with changes of habitat grads.

Gamma diversity

Whittaker recognised that the total diversity within a large area, which he called gamma diversity, could be partitioned into two components: local (alpha) diversity and turnover of species between habitats or localities (beta diversity). This partitioning of diversity is central to understanding the contribution of large-scale processes to local diversity because it elucidates the connection between local diversity and regional species richness.

Cody has suggested that in addition to turnover of species between habitats, species also replace one another over distance within a habitat type, a phenomenon he refers to as the gamma component of regional diversity. Cody's gamma diversity may reflect the allopatric distribution of closely related taxa; it may also express changes in physical environment within a given habitat type.

Whitaker's beta diversity includes the separate contributions that the habitat breadths of species and the variety of habitats in the region make to regional diversity. Alternatively, beta diversity may be thought of as an average property of species within a region and may be distinguished from contributions to regional diversity made by habitat variety and other dimensions of the total sample. Accordingly, beta diversity would be the inverse of the average number of localities or habitats occupied by each species within the larger region.

In this case, the contribution of beta diversity to regional diversity would be measured by multiplying it by the number of localities or habitats in the regional sample, that is, by the dimension of the sample. Thus principle, alpha diversity also can be partitioned into niche space components within the local sample. Each local assemblage of species occupies a certain niche volume, within which its total diversity equals the product of the average number of species occupying each point in niche space, the inverse of the average volume of niche space occupied by the entire assemblage.

Protecting Keystone Species

Within biological communities, certain species may be important in determining the ability of large numbers of other species to persist in the community. These crucial species have been termed keystone species. Protecting keystone species is a priority for conservation efforts, because if a keystone species is lost from a conservation area, numerous other species might be lost as well. Because many tropical insect species appear to be highly specialised in their feeding behaviour, subsisting on just one or a few related plant species, Raven has argued that the extinction cascade, with an additional loss of 10 to 30 insect species.

With the recent evidence of a greater diversity of insect species than previously suspected, it is even possible that the extinction of a single plant species could eliminate hundreds of insect species. Among the most obvious keystone species are top predators, since these species are often important in controlling herbivore populations. The elimination of even a small number of individual predators, constituting a minute amount of the community biomass, can result in dramatic changes in the vegetation and a great loss in biological diversity.

For example, in many localities where gray wolves have been hunted to extinction by humans, deer populations have exploded. The deer have then severely overgrazed the habitat, eliminating many herbaceous species. The loss of these plants was detrimental to the deer and to other herbivores, including the insect community that fed on the plants. In addition, the loss of plant cover has led to soil erosion, contributing to losses of the species that inhabit the soil. In short, the elimination of a keystone species-in this case, the wolves, which kept deer populations in check-can create a series of linked extinction events that result in a degraded ecosystem with much lower biological diversity at all trophic levels.

The importance of grazing animals in patterning communities is also illustrated by the case of a black sea urchin of the genus Diadema. This sea urchin was common in coral reefs throughout the tropical West Atlantic, where it fed on large, fleshy algae that were fairly uncommon. In 1981, following the unusually warm weather of an El Ni¨¾o event, there was a massive die-off of Diadema, apparently caused by a viral epidemic. Without the Diadema, the fleshy algae increased dramatically in abundance, overgrowing and damaging coral reefs.

The importance of a keystone species may hinge on highly specialised relationships between the keystone species and other organisms. In many tropical forests, fig trees and fig vines appear to be keystone species in the functioning of vertebrate communities. Fig flowers are pollinated by small, highly specialised fig wasps, which mature inside the developing fig fruit. Mature fig trees produce continuous fig crops, and generations of wasps are continually coming to maturity. As a consequence of this continuous fruit production, figs provide a reliable source of fruit to primates, birds and other fruit-eating vertebrates throughout the year, even during dry seasons.

While fig fruits do not have the high energy content of many preferred lipid-rich fruits or the high protein content of an insect diet, during drought periods the fig fruits serve as "famine food" that allows vertebrates to survive until their preferred foods are once more available. Even though fig trees and vines may be uncommon in the forest and the fruit may constitute only a small percentage of the total vertebrate diet, their persistence is necessary to the continued functioning of many species in the vertebrate community. In this case, the fig trees are a keystone species because so many other species rely on them for food, but the health of the Ficus population rests on the health of the wasp population; the relationship between the trees and the wasps forms the foundation of the entire community's health.

Many keystone species play less obvious roles that are nevertheless essential to maintaining biological diversity. Some inconspicuous detritivorous species play a significant role in the functioning of a community. For example, dung beetles exist at low density in tropical forests and constitute only a fraction of the biomass. Yet the beetles are crucial to the community because they bury dung and carrion as a food source for their larvae. These buried materials break down rapidly, making nutrients available for plant growth. Seeds contained in the dung of fruit-eating animals are also buried, which may facilitate seed germination and the establishment of new plants.

In addition, by burying and feeding on dung the beetles kill the parasites of vertebrates contained in the dung, helping to keep the vertebrate population healthy. The identification of keystone species has several important implications for conservation biology. First, as we have seen, the elimination of a keystone species from a community may precipitate the loss of many other species. Second, in order to protect a species of particular interest, such as a monkey, it may be necessary to protect the keystone species on which it depends either directly or indirectly, such as fig trees and dung beetles. Third, if the few keystone species of a community can be identified, these could be carefully protected or even encouraged if the area is being affected by human activity.

Habitat Fragmentation and Degradation

Alteration of habitats by human activity is the greatest threat to the richness of life on Earth. The most visible form of habitat alteration is direct habitat removal, as when a forest is clear-cut, a wetland is drained, a stream is dammed to create a reservoir, or a remnant prairie is converted to a shopping small. However, if we step back and view the broader landscape, as from a mountain peak or an airplane, the most striking pattern is often fragmentation of a once continuous natural landscape.

Habitat fragmentation has two components:

1. Reduction of the total amount of a habitat type, or perhaps of all natural habitat, in a landscape; and
2. Apportionment of the remaining habitat into smaller, more isolated patches.

Although the latter component is fragmentation in the literal sense, it usually occurs in tandem with widespread deforestation or other habitat reduction. In managed landscape, such as national forests, there are ways that vegetation can be removed without fragmenting the remaining vegetation. In some case, a landscape may be more "shredded" than fragmented. However, the end result of human settlement and resource extraction in a landscape is often a patchwork of small, isolated natural areas in a sea of developed land.

Studies in many regions have documented local extinctions, shifts in composition and abundance patterns to favour weedy species, and other forms of biotic impoverishment in fragmented landscapes. Thus, fragmentation has become a major subject of research and debate in conservation biology. Some effects of fragmentation on biodiversity have been conspicuous; others have been subtle and indirect. Some have occurred almost immediately after the initial disturbance, whereas others have developed over decades or are still unfolding.

Fragmentation of some regions, such as the Georgia Piedmont, has been partially reversed through abandonment of agriculture and maturation of second-growth forests. However, fragmentation of other regions, such as the Pacific Northwest, Florida, and much of the tropics, continues. Most deforestation in Central America has occurred since 1950 and is increasing in rate. The effects of fragmentation can be seen at several levels of biological organisation, from changes in gene frequencies within populations to continent-wide changes in distributions of species and ecosystems.

At the species level, there are essentially three options for persistence in a highly fragmented landscape. First, a species might survive or even thrive in the matrix of human land use; a number of weedy species worldwide fit this description. Second, a species might survive in a fragmented landscape by maintaining viable populations within individual habitat fragments; this is an cption only for species with small home ranges or otherwise modest area requirements, such as many plants and invertebrates. Many of these species can meet all of their life history requirements within the boundaries of a single fragment, barring major environmental change.

A third way to survive in a fragmented landscape is to be highly mobile. A mobile species might integrate a number of habitat patches, either into individual home ranges or into an inbreeding population. The Pileated Woodpecker has demonstrated adaptation to fragmented landscapes, particularly in eastern North America. Foraging individuals now travel among a number of small woodlots in landscapes that were formerly continuous forest, often using wooded fencerows as travel corridors. A species incapable of pursuing one or more of these three options is bound for eventual extinction in a fragmented landscape.

Barriers Isolation of Habitats

Isolation of habitats by movement barriers is an effect of fragmentation as important as reduction in habitat size. Species that are restricted to certain kinds of habitat may depend on a constellation of habitat patches in relatively close proximity, if no single patch is large enough to meet the needs of individuals or groups. The viability of metapopulations may depend on movement of individuals between patches being great enough to balance extirpation from local patches. Also, many animal species require a mix of different habitats with distinct resource-for example, food patches, roost sites, and breeding sites-in order to meet their life history requirements. If these critical areas become separated by barriers, populations may decline rapidly to extinction.

One significant type of barrier in many landscapes is roads. Habitat fragmentation is usually accompanied, and augmented, by road building. To the extent that individual animals hesitate to cross rods, roads fragment populations into smaller demographic units that are more vulnerable to extinction. A study in southeastern Ontario and Quebec found that several species of small mammals rarely ventured onto road surfaces when the road clearance exceeded 20 m.

In Oregon, dusk-footed woodrats (Neotoma fucipes) were trapped at all distances from an interstate highway right-of-way, but never in the right-off-way itself, suggesting that these rodents did not cross the highway. In Germany, several species of carabid beetles and two species of forest rodents rarely or never crossed two-lane roads; even a narrow, unpaved forest road, closed to public traffic, served as a barrier. Another study found that roads and railroads inhibited normal movements of lycosid spiders and carabid beetles; although crossings were rare, longitudinal movements along these barriers were stimulated.

Human-created barriers also fragment aquatic habitats. Dams, for example, not only block access of migratory fishes to upstream areas, but also prevent recolonisation of stream segments by any species whose local population have been extinguished due to natural or human causes. The flattened musk turtle has been lost from over half of its range in the Warrior River Basin of Alabama because habitat modification of stream and river channels has fragmented suitable habitat. The remaining populations are small and isolated, and therefore at high risk of extinction.

Local and Regional Extinctions

Metapopulation dynamics suggest that even once-common species are not immune to the effects of widespread habitat alteration and fragmentation. When local populations become isolated, they face a higher probability of extinction. For example, the Middle Spotted Woodpecker is a sedentary forest species with poor powers of dispersal. A population in Sweden, isolated since about 1950, remained relatively stable at 15-20 pairs from 1967-1974, and then declined rapidly to extinction in the period 1975-1983.

The proximate causes of extinction were mortality from cold weather and, more importantly, reproductive failures due to reduced fecundity, probably related to inbreeding depression. The White-backed Woodpecker has gone extinct in parts of Sweden where habitat fragmentation had resulted in a low density of suitable habitat patches; recolonisation of vacated patches in such areas is too low to maintain the metapopulation. Consideration of life histories suggests some hypotheses about relative vulnerability to fragmentation.

Among the categories of species predicted to be most vulnerable are the following: rare species, species with large home range, species with limited powers of dispersal, species with low reproductive potential, species dependent on resources that are unpredictable in time or space, ground-nesting species, species of habitat interior, and species exploited or persecuted by people. One of the most ominous effects of habitat fragmentation may not become apparent for decades or centuries: the inability of species in isolated fragments to track changes in habitat conditions related to changing climate.

Species Composition

Species composition and abundance patterns will be change in fragmented landscapes. For example, birds characteristic of forest interior habitats may be unable to maintain their populations in landscapes where edge is abundant; instead, the; the landscape may gradually become dominated by edge-adapted species not in great need of conservation. Species composition is altered in fragmented landscapes because some species are more vulnerable than others to reduced area, increased isolation, edge effects, and other factors that accompany the fragmentation process. Species loss from fragmented habitats, then may follow a predictable and deterministic sequence.

In studying habitat patches of various size, a pattern of nest subsets in the distribution of species is often observed. A nested subset is a geographic pattern in which larger habitats contain the same subset of species found in smaller habitats, but add new species to that subset. Thus, a set of the most common species is found in all habitat sizes from smallest to largest, but progressively larger habitats add subsets of species found only in larger areas. Boreal mammals and birds on mountain ranges in North America's Great Basin, which are natural habitat islands, show a nested distribution pattern that may be a consequence of selective extinction of area-dependent species on smaller islands.

In this case, extinctions occurred in the same basic sequence throughout the region, despite considerable variation in extinction rates. Distribution of bird species among woodlots in agricultural landscapes is typically nonrandom; species found in small woodlots are also found in the larger sites. In east-central Illinois, the most highly nested pattern was found for species requiring forest interior habitat for nesting and for species that migrate to the Neotropics. A similar nonrandom pattern, with all species occupying large patches but many species not occurring in small patches, has been documented for birds in Swedish lakes and many other temperate communities.

Studies of birds in the Wheatbelt of Western Australia documented loss of many species in small habitat remnants since isolation, a relaxation effect predictable from island biogeographic theory. Such results support previous suggestions that although a collection of small sites may harbor more species, large sites are needed to maintain populations of species sensitive to human disturbance. Recognising that "conservation strategy should not treat all species as equal but must focus on species and habitats threatened by human activities", an optimal conservation strategy in most cases would avoid fragmentation of large natural areas.

The addition of species to fragmented landscapes is as important to consider in conservation strategy as is species loss. Many of the species that invade fragmented landscapes are exotics. Increases in the number of species as a local scale due to invasions by weedy or exotic species are often accompanied by declines in diversity at a regional or global scale as sensitive native species are progressively lost, even though overall species invade more and more regions, regional biotas are homogenised and lose their distinctness—a "mongrelisation" of the global landscape. This process of homogenisation is one of the most prominent forms of biotic impoverishment worldwide.

Fragmentation generally increases the rate of exotic species invasion, often through creation of disturbed habitats through which exotics travel rapidly. For example, many fungal diseases of trees, exotic plants, and exotic insect pests are known to disperse and invade natural habitats in the western United States via roads and vehicles. Sometimes habitat fragments can be relatively resistant to invasion. Patches of old-growth forest in Indiana had alien plants along their edges, but forest interiors were virtually free of aliens. The dense wall of vegetation that developed along the edges of fragments in this study was thought to discourage invasion by reducing interior light levels and wind speeds.

Climate change

Species migrating in response to climate change have always had to cope with dispersal barriers such as rivers, lakes, mountain ranges, and desert basins. The additional set of barriers created by human activities will make migration all the more difficult. New climates will render reserves set aside to protect certain species or communities unsuitable for them. Weeds may dominate many fragments. Even natural rates of climate change threaten species restricted to fragments surrounded by inhospitable habitat.

The increased rates of change predicted with greenhouse warming may eliminate all but the most vagile species as they fail to track shifting climatic conditions. High-elevation and high-latitude habitat may be lost entirely. Although wide habitat corridors and artificial translocations of populations northward and upslope may help some species in some areas, these solutions will not suffice for whole communities, especially if climate change is as rapid as predicted.

In sum, fragmentation, the loss and isolation of natural habitats, is one of the greatest threats to regional and global biodiversity. Whereas natural disturbances and other processes create heterogeneous landscapes rich in native species, human land uses often create islands of natural habitat embedded in a hostile matrix. Such fragmentation reduces or prevents normal dispersal, which is critical to long-term population viability for many species, and increases edge effects and other threats.

COMMUNITY-LEVEL CONSERVATION

By definition, conservation of biodiversity must ultimately reach beyond the species level and consider communities and ecosystems. Knowledge of a community's particularly critical components and strong interactions such as keystone species and mutualisms, can focus and guide conservation actions. A species may be a keystone component typically because of some aspect of its trophic ecology, such as being an important predator, changing habitat structure through foraging efforts, or providing a critical food resource.

Mutualisms create critical links among species, such as pollination or seed dispersal that perform a function that may be critical to continued community composition. Changes occur in communities over ecological time for a variety of reasons, but especially relevant to conservation are natural disturbance regimes and the introduction of exotic species. Virtually all natural communities experience a normal disturbance regime, such as fire, flooding, drought, or storm damage. Understanding the scale, frequency, and intensity of natural disturbances is critical to conservation at community level.

Continuance of normal disturbance regimes at appropriate scales should be encouraged, and new, artificial disturbance is invasive species. Many examples exist of the destructive effects of both plant and animal invasions. Their success depends on the characteristics of both the particular invader and the particular communities invaded, and is difficult to predict a priori. All species introductions to and losses from communities are not equal with respect to their community effects; some may be quit benign, while others can devastate. Long-term studies of ecological communities over evolutionary and geological time shed more light on community conservation.

Many communities especially in temperate zone have been dynamic over spans of thousands of years, and some may be relatively new communities in evolutionary time. Species additions to or deletions from communities in evolutionary time are often independent; that is, many communities do not seem to be cohesive, tightly woven units that remain together over time, but assemblages of species that continually enter and leave the picture.

These types of communities may be less susceptible to human disturbance than are communities with long-term associations that have evolved complicated species interactions. Thus, the changes wrought by humans on natural communities are superimposed upon a historical dynamism in many cases, making conservation more difficult, but providing another perspective on ecological change that should be incorporated into management efforts.

Fragmentation acts to reduce biodiversity through four major mechanisms. First, because remaining fragments represent only a sample of the original habitat, many species will be eliminated by chance. Second, the modified landscape in which fragments exist is often inhospitable to many native species, thus preventing normal movements and dispersal. Third, small fragments contain fewer habitats, support smaller populations of native species that are more susceptible to extinction, and are less likely to intercept paths of dispersing individuals. Finally, climatic influences and opportunistic predators and competitors from the disturbed landscape penetrate into fragments, reducing the core area of suitable habitat.

Many terrestrial game animals in North America are adapted to habitat edges and thrive in fragmented landscapes, which has encouraged wildlife managers to promote edge habitat at the expense of sensitive species of habitat interiors. Some strategies for countering fragmentation are available:

1. Do not fragment intact wildlands and other natural areas, few of which remain in most regions;
2. Minimise road construction, clearing of vegetation, and creation of other barriers to dispersal;

3. Maintain or restore wide habitat corridors or other forms of functional connectivity between natural area;
4. Minimise creation of artificial dispersal corridors that encourage proliferation of exotic species.

TYPES OF ECOSYSTEMS

Attempts to classify large ecological units consisting of all of the species living together in a region have a long and complicated history. Plants have been used extensively in terrestrial ecological classification systems because they dominate the structure of those communities and are much easier to sample than are animals. Alexander von Humboldt classified vegetation into groups on the basis of the shapes of the dominant plants, such as spruce-tree shape or palm shape.

In 1874, de Candolle proposed a classification of plant communities based on life-forms because he believed that life-forms were determined by climate. More complicated schemes of describing life-forms of plants and characterising plant communities by the relative dominate of plants with different shapes have been developed. Although ecological communities clearly grade into one another, recognition of major divisions has proven to be useful for ecological discourse. Most textbooks divide the world into biomes, the largest ecological units, on the basis of the dominant vegetation.

In tropical regions, biomes are divided along precipitation gradients and elevational gradients. At temperate latitudes, commonly recognised biomes along a moisture gradient are mesophytic forest, woodland, tallgrass prairie, shortgrass prairie, and desert. With increasing latitude, mesophytic forests become conifer-dominated boreal forests and eventually tundra. Finer divisions of biomes into ecosystem types use the same types of information, but also use data on drainage, soil type, slope, and species composition to identify units.

All such classifications are arbitrary, but if properly constituted, they identify ecological units that are useful for comparative purposes. Preservation of the variety of the earth's biomes is necessary for preservation of species. Without sufficient quantities of their natural habitats, species become extinct in the wild. Captive propagation can, and does, serve a role in keeping species alive for short periods until they can be reintroduced into the wild. But captive propagation is of little ultimate use if there are no suitable sites into which to reintroduce the species.

Managing species in zoos and botanical gardens is expensive, and an animal in a cage or a plant in a garden in not a fully functioning member of its species. Also, ecosystems can provide the environmental goods and services upon which human life depends only if well-functioning systems are distributed over most of the surface of the earth. This requires a rich variety of ecosystems because the combinations of species that can perform as an integrated ecosystem change with environmental conditions. Deciduous trees cannot adequately replace conifers in boreal environments, nor can conifers richness of our aesthetic experiences depends upon the richness of biomes as well as the richness of species.

Arid and Semi-arid Zone

Grasslands and savannahs: Grassland ecosystems are areas dominated by grasses, grass-like plants and woody plants and are further characterised by periodic drought, fire and large herbivore grazing. Soils are usually marked by low fertility. In particular, savannahs are characterised by a dominant grassy ground layer which can be present in treeless plains, open woodlands, and closed-canopy woodlands. The earth's land surface is composed of 20 per cent grasslands with temperate grasslands representing one forth of the area and savannas the remainder.

Warm desert and semi-desert: Hyper-arid desert ecosystems have irregular rainfall with periodic droughts lasting several years. The Sahara desert represents nearly 70 per cent of the global hyper-arid total. Semi-desert areas are dominated by succulents and other xerophytic plants and are often characterised by sparse or absent tree cover. Rainfall in arid and semi-arid areas varies from less than 200 to a maximum of 800 mm annually.

Tundra communities and cold desert biomes: Non-humid areas are composed of cold regions and drylands and represent 61 per cent of the Earth's total land surface. Cold regions include polar and tundra areas as well as certain high mountains and plateaus and compose 14 per cent of the non-humid area total. Temperatures remain below freezing for a period long enough to restrict plant growth and liquid water is usually unavailable for a significant part of the year.

Marine and Freshwater Ecosystems

Mangroves: A hybrid terrestrial/marine ecosystem, mangroves are composed of shrubs, trees, and ferns living in or adjacent to the intertidal zone. Mangroves are predominantly tropical and are rarely found beyond the latitudes 30°N and 30°S with the most diverse systems located in Southeast Asia.

Coral reefs: Composed of calcium carbonate, corals are most abundant in shallow marine environments usually characterised by warm low-nutrient waters. Coral reefs are among the most diverse ecosystems in the world with over 450 different coral species identified in Southeast Asia alone.

Seagrasses: Found in shallow coastal areas between the Arctic and Antarctic, seagrasses are flowering plants that live submerged in seawater. Entire ecosystems are founded on their high productivity as they provide an important source of food for many organisms.

Pelagic or open-ocean communities: Founded on the activity of plankton, which supports a large number of free-swimming organisms, pelagic communities occupy a greater area than any other major community on earth. Characterised by vertical gradients that fluctuate seasonally or as often as daily, the community responds to variations in the physical and chemical variations of seawater.

Deep-sea communities: Deep-sea communities exist in permanent darkness at depths greater than 3000 metres. Such areas are marked by low biomass, although sampling techniques indicate small organism diversity near the sea bottom is relatively high.

Rivers: Bodies of freshwater draining from elevated land toward sea level, rivers show considerable variation in water flow, underlying geology, altitude and latitude. While there can be a wide range of habitats within an individual river, some rivers are marked by continual change while others have been in continuous existence for millions of years.

Lakes: Lakes are mainly glacial in origin. As a rule, they are geologically short-lived, typically changing or disappearing within 100,000 years. As a result, lake ecosystems are often less complex—less evolved—than others.

Wetlands: Wetlands are characterised by permanent or seasonal shallow waters and large aquatic plants and are usually categorised into one of four major types: bogs, fens, marshes, and swamps. Bogs are peat-producing wetlands in most climates in which organic matter has accumulated over long periods. Fens, also peat-producing wetlands, are usually supplied by mineral-rich groundwater. Marshes and swamps are inundated areas with herbaceous emergent vegetation or forested wetlands where little or no peat production occurs.

Forest Ecosystems

Boreal and temperate needleleaf forests: This forest type covers a larger area than any other. Species diversity is usually relatively low and dominated by a small number of tree species. However, diversity among organisms such as lichens and mosses is usually relatively high. Serving as major reservoirs of organic carbon, these forests are significant in the carbon cycle both above and below ground.

Temperate broadleaf and mixed forests: Structurally more complex than needleleaf forests with considerably higher species diversity, the temperate forests are generally representative of warmer latitudes. Aboveground biomass is lower than that of temperate needleleaf with below ground biomass being intermediate.

Tropical moist forests: The most diverse ecosystems on earth, tropical moist forests contain an estimated 60 per cent of all species. Diversity is extremely high with as many as 300 tree species per hectare in areas such as the western Amazon. Such forests account for nearly one third of the annual global terrestrial net primary production.

Tropical dry forests: Characteristic of tropical areas affected by seasonal drought, tropical dry forests are often composed of a deciduous forest canopy. Sparse trees and parkland occupy areas in transition from forested to non-forested landscapes and most tropical dry forests have lower species diversity than tropical moist forests.

Forest plantations: Forest plantations were usually intended for the production of timber and pulpwood and often consist of only one species. Not usually important as a habitat for native species, such systems can maintain nutrient capital and soil structure as well as serve to protect watersheds.

Mountain Ecosystems

Mountains and highlands represent one quarter of the earth's land surface. Mountain ecosystems are susceptible to change from such environmental influences as landslides, avalanches, lava flows and torrents and are slow to recover. Recently, climate change has led to the rapid melting of glaciers and icecaps in many mountain ranges. Mountain ecosystems are often highly biodiverse, with large numbers of endemic species.

Impact of Human Activity

The Earth has limited resources to support the organisms that live on it. Increasing human population numbers are putting great pressure on many of these limited resources and deplete those resources which can not be renewed. Many different natural processes occur within those ecosystems influencing humans. Some of these processes include atmospheric quality. soil generation and conservation, energy flow, the water cycle, waste removal and recycling.

Human activities are altering the equilibrium involved in these natural processes and cycles. If these changes due to human activities are not addressed, the stability of the world's ecosystems may irreversibly affected. Human activities which have harmed ecosystems have resulted in a loss of diversity in both living things and the nonliving environment. Examples of these changes include land use, the cutting of vast areas of forest, and pollution of the soil, air, and water. Another way humans have changed ecosystems in a harmful way is by adding or removing specific organisms to these ecosystems.

Our ever increasing demand for energy has impacted ecosystems negatively as well. Many environmental risks are associated with our use of fossil and nuclear fuels. Many factors associated with human populations have influenced environmental quality. These include population growth and distribution on our planet, our use of resources, the ability of technology to solve environmental problems, as well as the role of economic, political, ethical, and cultural views in solving these problems.

Individual choices and the actions of society can contribute to the improvement of our environmental problems. Our choices which must include an assessment of the risks, costs, benefits, and trade-offs of new technologies and continued human expansion. All changes and proposed improvements need to consider both the human and environmental impact of the change.

Natural Environment

Natural environment is of crucial importance for social and economic life. We use the living world as:

1. A resource for food supply,
2. An energy source,
3. A source for recreation,
4. A major source of medicines,
5. Natural resources for industrial products.

In this respect the diversity of nature not only offers man a vast power of choice for his current needs and desires. It also enhances the role of nature as a source of solutions for the future needs and challenges of mankind.

State of Ecosystems and Species

In the past, human interaction with nature, although often having a disruptive effect on nature, often also enriched the quality and variety of the living world and its habitats—e.g. through the creation of artificial landscapes and soil cultivation by local farmers.

Today, however, human pressure on natural environments is greater than before in terms of magnitude and efficiency in disrupting nature and natural landscapes, most notably:

1. Intensive agriculture replacing traditional farming; this combined with the subsidies of industrial farming has had an enormous effect on western rural landscapes and continues to be a threat.
2. Mass tourism affecting mountains and coasts.
3. The policies pursued in the industry, transport and energy sectors having a direct and damaging impact on the coasts, major rivers (dam construction and associated canal building) and mountain landscapes (main road networks).
4. The strong focus of forestry management on economic targets primarily causes the decline in biodiversity, soil erosion and other related effects.

The consequences are:

1. A decreased species diversity, due to reduced habitable surface area which corresponds to a reduced "species carrying capacity".
2. The reduction of the size of habitats also reduces the genetic diversity of the species living there. Smaller habitats can only accommodate smaller populations, this results in an impoverished gene pool.
3. The reduction of genetic resources of a species diminishes its flexibility and evolutionary

adaptability to changing situations. This has significant negative impacts on its survival.

The conditions under which the reduction of habitats often occur prevent living organisms making use of their normal ways to flee their threatened habitat. Those escape routes include migration to other habitats, adaption to the changing environment, or genetic interchange with populations in nearby habitats. Of particular concern is:

1. The abrupt nature of human intervention; human projects are planned and implemented on a much shorter time scale than natural processes;
2. Furthermore human intervention, such as the construction of buildings, motorways or railways results in the fragmentation of habitats, which strongly limits the possibility for contact or migration among them;
3. In extreme cases even the smallest, narrowest connections between habitats are broken off. Such isolation is catastrophic for life in the habitat fragments.
4. *Loss of Species of Fauna and Flora*: Although relatively few species of Europe's fauna and flora have actually become totally extinct during this century, the continent's biodiversity is affected by decreasing species numbers and the loss of habitats in many regions. Approximately 30% of the vertebrates and 20% of the higher plants are classified as "threatened". Threats are directly linked to the loss of habitats due to destruction, modification and fragmentation of ecosystems as well as from overuse of pesticides and herbicides, intensive farming methods, hunting and general human disturbance. The overall deterioration of Europe's air and water quality add to the detrimental influence.

3

NATURAL RESOURCES

A natural resource's value rests in the amount of the material available and the demand for it. The latter is determined by its usefulness to production. A commodity is generally considered a natural resource when the primary activities associated with it are extraction and purification, as opposed to creation. Thus, mining, petroleum extraction, fishing, hunting, and forestry are generally considered natural-resource industries, while agriculture is not.

CLASSIFICATION OF NATURAL RESOURCES

Natural resources are mostly classified into renewable and non-renewable resources.

Renewable Resources

A natural resource qualifies as a renewable resource if it is replenished by natural processes at a rate comparable to its rate of consumption by humans or other users. Resources such as solar radiation, tides, and winds are *perpetual resources* that are in no danger of being used in excess of their long-term availability.

Natural resources that qualify as renewable resources include oxygen, fresh water, timber, and biomass.

However they can become non-renewable resources if used at a rate greater than the environment's capacity to replenish them. For example, groundwater may be removed from an aquifer at a rate greater than the sustainable recharge. Removal of water from the pore spaces may cause permanent compaction that cannot be reversed.

Renewable resources may also include commodities such as wood, paper, and leather. Furthermore alcohol is also a renewable source of energy, similarly, oils from plants and seeds can used as even as a substitute for non-renewable diesel, last but not least methane is also considered as a renewabie source of energy. Gasoline, coal, natural gas, diesel and other commodities derived from fossil fuels are non-renewable. Unlike fossil fuels, a renewable resource can have a sustainable yield.

Wind Power

Wind power is the conversion of wind energy into more useful forms, usually electricity using wind turbines. In 2005, worldwide capacity of wind-powered generators was 58,982 megawatts; although it currently produces less than 1 per cent of world-wide electricity use, it accounts for approximately 23 per cent of electricity use in Denmark, 9 per cent kane, and 6 per cent in Germany. Globally, wind power generation less than quadrupled between 1999 and 2005. Most modern wind power was developed in 2003 by Wally Watson which generates wind in the form of electricity without smoke converting the rotation of the turbine blades into electrical current by means of a generator.

In windmills wind energy is used to turn mechanical machinery to do physical work, like crushing grain or pumping water. Wind power is used in large scale wind farms for national electrical grids as well as in small individual turbines for providing electricity to rural residences or grid-isolated locations. Wind energy is ample, renewable, widely distributed, clean, and mitigates the greenhouse effect if used to replace fossil-fuel-derived lightning.

To make the power, the windmill captures the wind and spins a generator. The generator then gives off gas which then forms into electricity. The siting of turbines has become a controversial issue amongst those concerned about the value of natural landscapes, paricularly since the best sites for wind generation tend to be in scenic mountain and oceaside areas.

Hydro-power

Hydro-power is the capture of the energy of moving water for some useful purpose. Prior to the widespread availability of commercial electric power, hydropower was used for irrigation, milling of grain, textile manufacture, and the operation of sawmills. The energy of moving water has been used for centuries; in Imperial Rome, water powered mills produced flour from grain, and in China and the rest of the Far East, hydraulically operated "pot wheel" pumps raised water into irrigation canals.

In the 1830s, at the peak of the canal-building era, hydropower was used to transport barge traffic up and down steep hills using inclined plane railroads. Direct mechanical power transmission required that industries using hydropower had to be situated near the waterfall.

For example, during the last half of the 19th century, many grist mills were built at Saint Anthony Falls, utilising the 50 foot (15 metre) drop in the Mississippi River. The mills contributed to the growth of Minneapolis. Today the largest use of hydro-power is for electric power generation, which allows low cost energy to be used at long distances from the watercourse.

Non-renewable Resources

A non-renewable resource is a natural resource that cannot be re-made, re-grown or regenerated on a scale comparative to its consumption. It exists in a fixed amount that is being renewed or is used up faster than it can be made by nature. Fossil fuels and nuclear power are non-renewable resources, as they do not naturally re-form at a rate that makes the way we use them sustainable and consumer materials to produce electricity.

A renewable resource differs in that it may be used but not used up. This is as opposed to natural resources such as timber, which re-grows naturally and can, in theory, be harvested sustainably at a constant rate without depleting the existing resource pool and resources such as metals, which, although they are not replenished, are not destroyed when used and can be recycled. A non-renewable resource is always drawn down with anabolic processes that use up energy.

Carbon-based Resources

Carbon-based resources such as coal, oil, or natural gas, take millions of years to form naturally and cannot be replaced as fast as they are consumed. Eventually they will be used up. At present, the main energy sources used by humans are non-renewable; renewable resources, such as solar, tidal, wind, and geothermal are yet to be fully utilised. Fossil fuels like coal, oil, and gas generate a considerable amount of energy when they are burnt.

Non-renewable resources have a high carbon content because their origin lies in the photosynthetic activity of plants millions of years ago. The fuels release this carbon back into the atmosphere as carbon dioxide. The rate at hich such fuels are being burnt is thus resulting in a se in the concentration of carbon dioxide in the tmosphere, a cause of the greenhouse effect. The sun can lso be a nonrenewable resource in some ways.

Natural resources are replaced by natural processes iven unreasonable amount of time. Soil, water, forests, lants, and animals are all renewable resources as long as ey are properly conserved. Solar, wind, wave, and eothermal energies are based on renewable resources. lowever, non-renewable resources may take billions of ears to form. Renewable resources such as water, wind ower, sun, etc are practically infinite—they cannot be epleted—unlike their non-renewable counterparts, hich will have run out by 2200.

apital Processes

atural resources are converted to commodity inputs to frastructural capital processes. They include soil, mber, oil, minerals, and other goods taken more or less om the Earth. Both extraction of the basic resource and fining it into a purer, directly usable form, are generally onsidered natural-resource activities, even though the tter may not necessarily occur near the former. A ation's natural resources often determine its wealth and atus in the world economic system, by determining its olitical influence in.

Developed nations are those which are less dependent n natural resources for wealth, due to their greater liance on infrastructural capital for production. owever, some see a resource curse whereby easily otainable natural resources could actually hurt the ospects of a national economy by fostering political rruption. In recent years, the depletion of natural pital and attempts to move to sustainable development ive been a major focus of development agencies.

This is of particular concern in rainforest regions which hold most of the Earth's natural biodiversity— irreplaceable genetic natural capital. Conservation o natural resources is the major focus of natural capitalism environmentalism, the ecology movement, and Gree Parties. Some view this depletion as a major source o social unrest and conflicts in developing nations.

RESOURCE MANAGEMENT AND CONSERVATION

As for wildlife, conservation science falls into two basi categories:

1. Threat assessment, and
2. The analysis of small populations of animals.

The two basic categories can as well be applied to natura resources, for there is a threat to these resources whic has to be assessed and there is need to analyse the smal limited amounts of resources available for use and t examine the possibilities for conservation. One of th results of the two activities above was the creation c various kinds of protected areas. The focus of th conservation community has been on setting asid ecologically important marine and terrestrial areas reducing the over-harvesting of wildlife, lessening th pollution of fragile lands and waters, and protecting lon term ecological processes.

Also fitful cooperation between scientists and inte governmental organisations have resulted in som important conservation achievements over the year namely, the Ramsar Convention on Wetlands, th Atmospheric Nuclear Test Ban, Bio-safety Protocol, th Convention on International Trade in Endangered Speci of Wild Fauna and Flora and the Creation of Worl Heritage Sites. But these have not added up to gre successes, rather a tenuous grip on the global politic agenda.

The blame for little success must rest with the developing as well as the developed countries. The former have shied away from their post-Rio obligations but they do contribute to much future growth in fossil-fuel consumption and natural resources use. The latter have been going out of their to capture and use world's natural resources for their own economic ends, knowing that diminishing returns have been operative for long. Besides, they use natural resources far in excess of their needs and in the bargain deprive the developing countries their right to use their own. Having said these words, let us now turn to conservation and preservation and look at them in a historical way.

History

The idea that humans have abused the planet's natural resources and environment is not modern in origin. The earliest writings suggesting concern for the environment date back to the time of the ancient Greeks, when Plato described the effects of soil degradation and deforestation on the peninsula of Attica. This brief account was followed by the accounts of many others as they documented the negative effects of agriculture, land use change, urbanisation, and industrialisation.

In the first century, Columella and Pliny the Elder warned that poor land management in the croplands around Rome could have a negative effect on crop yields and may cause soil erosion. Throughout the middle ages, countless accounts describe the effects of pollution and degradation on the environment. Overpopulation and the subsequent resource depletion are believed to be the cause of the collapse of the Mayan Civilisation in the tenth century.

During the middle ages, large tracts of forests were removed for their wood or converted into pasture and cropland in much of Europe and Asia. One of the fir examples of deliberate preservation of wildlife occurre during this period when noblemen in Britain and Europ put aside areas of land for the purpose of hunting. In th mid-1700s, the Industrial Revolution began in Englan During this period most of the forests had been cu down, and coal was used to replace wood as source energy.

Burning coal produced atmospheric pollution on local and regional scale. John Evelyn, a naturalis complained in 1661 of the effects of atmospheric pollutio on the air quality in London. The Environment Movement in Britain during the sixteenth, seventeent and eighteenth centuries was primarily influenced b three major forces. Scientific discovery by naturalis provided a general understanding of how nature work It also allowed for the recognition that huma degradation of the environment influenced the surviv of living organisms.

By the 1880s, field and naturalist clubs had combined membership of over 100,000 people. At th time, many of the members of these organisations we active in building collections of native birds, eggs an plants. However, the combined effort of the collecto was having a drastic effect on species numbers. As result of the declining numbers of collected species, wildlife protection movement developed. The last maj force was the growing reaction to the environment degradation caused by urbanisation. Urbanisatio resulted in air pollution, water pollution, and th conversion of natural space into built environments.

The first significant developments in enviro mentalism and conservation took place in the 19t century. In 1847, George Perkins Marsh gave a speech the Agricultural Society of Rutland County, Vermont. Th subject of this speech was that human activity wa having a destructive impact on land, especially through deforestation and land conversion. The speech also became the foundation for his book Man and Nature or The Earth as Modified by Human Action, first published in 1864. In this book, Marsh warned of the ecological consequences of the continued development of the frontier.

Henry David Thoreau wrote the famous book about conservation and the environment called Walden in 1854. In the book Walden expressed the idea that human civilization was becoming too complex and removed from its foundations in the natural world. Thoreau suggested that humanity should simplify its economic and societal systems so that they are more in harmony with nature. He also suggested that humans should strive for environmental wisdom—which is the ability to make correct decisions and long-term planning by sorting through natural and human created facts and information.

In 1892, the Sierra Club was incorporated in the United States with John Muir as President. John Muir suggested that the utilitarian approach to resource management did not go far enough to protect nature. He suggested that certain resources should be permanently preserved and protected. As a result of his views, some areas in the United States were put aside as National Parks. Influenced by Marsh's book, President Theodore Roosevelt decided to change the way the United States Government managed natural resources.

With the help of Gifford Pinchot, his main conservation advisor, policies were developed and laws were passed to insure that resources would be managed using utilitarian principles. Utilitarian conservation suggests that renewable resources should be managed so that they will never be exhausted. President Theodore Roosevelt creates first national wildlife refuge, on Pelican

Island, Florida. By 1909, the Roosevelt Government creates 200,000 square kilometres of national forests, 51 national wildlife refuges and 18 areas of special interest, including the Grand Canyon. In the mid-20th century, the Environmental Movement began growing at an accelerating rate as human misuse of the Earth's environment increased and became more visible to the public.

Water Conservation and Resources Management

Human demand for water has been growing for two reasons. Firstly, it is primarily because of the growth of the human population. Secondly, as the human standards of living improve in the various countries of the world, the demand for water also increases to meet new needs in industry, agriculture and domestic use. However, there is a limited quantity of fresh water on Earth and for many nations this resource is scarce in its availability. People in Canada and the United States, for example, believe that clean water is available to them in an unlimited supply.

Nevertheless, supplies are not unlimited but finite and increasing demand for this resource will soon create problems that can only be corrected by management and conservation. People are also beginning to recognise that water is important for things other than domestic, agricultural and industrial purposes. Water is important for maintaining fish and wildlife populations, for recreation and for aesthetics. Governments in many countries have now established water resource management programmes that aim to provide a sustainable supply of high quality water in an efficient and environmentally sound manner.

Conservation Techniques

A number of techniques and technologies can be used to

make agricultural, industrial and domestic water use more efficient.

Reductions can easily occur in the following areas:

Reducing Agricultural Waste: Irrigation accounts for 70 per cent of the world's water use. Most irrigation systems deliver water to crops by flooding the land surface, diverting water to fields via open channels, or by sprinkler systems that apply water to the field surface. In general, these methods are very inefficient as only 50 per cent of the water applied is absorbed by the plants. The rest is lost to the atmosphere by evaporation. Micro-irrigation techniques can reduce the amount of water applied to crops by 40 to 60 per cent.

Other strategies that can be used to reduce agricultural water use include:

1. The cultivation of food crops that require less water for growth.
2. The use of lined or covered irrigation canals to reduce infiltration and evaporation losses.
3. Irrigating crops at night or early morning when evaporation potentials are low.
4. Reduce water subsidies and encourage the proper pricing of this resource.

Reducing Industrial Waste: Industry is the second largest user of water supplies. Reducing the amount of water used in industry not only makes more water available for other purposes but it can also reduce the volume of pollution. Industry reductions can be achieved by:

Designing industrial processes to recycle water: For example, water used for industrial cooling purposes can be cooled down in a cooling tower and then reused.

— Increasing the cost of water to industries to encourage water recycling.

- Recycling materials themselves can also greatly reduce water demand. For example, manufacturing a ton of aluminum from scrap rather than from virgin ore can reduce the volume of water used by 97 per cent.

Reducing Domestic Waste: Some strategies for reducing domestic consumption include:

- Replace lawns in semiarid and arid urban areas with xeriscaped surfaces.
- Encourage the use of efficient irrigation systems for home garden and lawn use.
- Manufacture and legislate the use of more efficient dishwashers, washing machines, and bathroom showers and toilets.
- Encourage leak detection and repair for distribution systems. Distribution systems in many of the world's urban areas are losing between 25 and 50 per cent of their water supplies due to leaks in pipes.
- Properly price water for domestic use. This price must reflect the environmental cost of over consumption and resource degradation. Many studies have shown that higher prices for water provide motivation for people to conserve. The introduction of water meters in Boulder, Colorado reduced water use by about 30 per cent. In Canada, water is metered in approximately two-thirds of the municipalities.
- Education can encourage people to reduce the amount of personal consumption.

Methods to Increasing Water Supplies

Humans have used several different methods to increase supplies of water. Some of these techniques involve the modification of the runoff process. Dams and reservoirs have been used for many centuries to trap runoff behind earth or concrete walls. The stored water is then transferred via canals or aqueducts for use in agriculture, industry, or domestic processes. Worldwide there are now over 36,000 operational dams, some of which are also used to generate energy. Several problems can occur with the storage of water in these human created features.

In some reservoirs, sediments can accumulate to a point where they can no longer be used for water storage or hydroelectric production. Other reservoirs have severe evaporation or leakage problems. Large amounts of water are annually lost from the Aswan High Dam in Egypt because of evaporation. This problem has reduced the planned amount of irrigation water supplied by this dam by one-half. In recent years, many nations have increased their supply of freshwater by exploiting the water found beneath the Earth's surface.

Groundwater contains more than 10 per cent of the freshwater found in the hydrosphere. Saudi Arabia receives 75 per cent of its water supply from groundwater mining. In many cases, withdrawal rates of this water greatly exceed the natural rates of recharge. Depleting groundwater reservoirs can lead to a number of problems, including: subsidence, earthquakes, sinkhole development, and saltwater intrusion.

Many projects have used canals, aqueducts, and diversion techniques to move water to places of need. In the former Soviet Union, diversions on the Amu Dar'ya and Syr Dar'ya Rivers have been used to create irrigation water for cropland. However, these diversions are also responsible for reducing the flow of runoff water to the Aral Sea. Because of the reduced flow, the Aral Sea has declined in area by over 50 per cent, has lost two-thirds of its volume, and has greatly increased in salinity. At current rates of reduction, the Aral Sea could be gone by 2020.

Rainwater Harvesting

Rainwater harvesting may be defined as process of augmenting the natural infiltration of rainwater or surface run off into the ground by some artificial methods. The methods suggested are recharge through pits, trenches, bore well shafts by directly diverting run off water into existing or disused wells or conserving the rain water by artificial storing and using the same for human use. The choice and effectiveness of any particular method is governed by local hydrological and soil conditions and ultimate use of water.

Need for Rainwater Harvesting: Nature replenishes the ground water resources annually through rainfall; by way of infiltration though soil layers. In urban areas, due to urbanisation, the soil surface exposed to natural recharge gets reduced. Therefore, natural recharge is diminishing, resulting in drying of wells. Groundwater source has the benefit of availability where water is needed and during emergencies and scarcity period, the public at large or NGOs should take measures to improve the groundwater recharge by rain water harvesting to maintain reliable and sustainable groundwater resources.

Rainwater Harvests and Techniques: For many countries, particularly those with monsoonal climates and long dry seasons, water shortages result not from a lack of rainfall but from a seasonally uneven supply. When annual rainfall is concentrated in a few months, storage is difficult. To illustrate, India has 2.1 trillion cubic metres of freshwater available each year, and the United States has 2.5 trillion cubic metres. While rain falls in the United States throughout the year, in India, which is geographically only one third as large, most of the rainfall comes between mid-June and mid-September.

As a result, most of this deluge runs off and is quickly carried back to the sea by the country's rivers. Although there are thousands of dams in India, they can collectively store only a fraction of the rainfall. The focus on building large dams to capture and store surface water before it runs off dominated most of the last century. But because sites were becoming scarce and because the construction of large dams often inundates large areas, displacing local populations and irreversibly altering local ecosystems, this era has now largely run its course. More and more countries are turning to local water harvesting to ensure adequate supply.

In Rajasthan, where water shortages constrain development and prevent people from escaping poverty, the villagers under the leadership of a movement helped design local water storage facilities. Once villagers helped select a site, they would organise to build an earthen dam. All the materials, the stone and the earth, were local. So too was the labour—sweat equity provided by the villagers. The leader would help with the engineering and design. He told villagers that in addition to meeting their daily needs for water, the seepage from the small reservoir would gradually raise the water table, restoring wells that had been abandoned. He also told them this would take time. It worked exactly as he said it would.

The initial success led to the creation of a local non-governmental organisation with 45 full-time employees and 230 part-timers. Funded by the Ford Foundation and other groups, it has not only helped build 4,500 local water storage structures in Rajasthan, it has also raised villagers' incomes and improved their lives. When the local topography is favourable for building successful small water storage structures, this can be a boon for local communities. This approach works not only in monsoonal climates, but also in arid regions where low rainfall is retained for local use. With a modest amount of engineering guidance, hundreds of thousands of communities worldwide can build water storage works.

Another technique to retain rainfall is the construction of ridge terraces on hillsides to trap rainfall near where it falls, letting it soak into the soil rather than run off. Using a plow to establish the ridges, local farmers can build these terraces on their own, but they are more successful if they are guided by a surveyor who helps establish the ridgelines and determines how far apart the ridges or terraces should be on the hill. Once the terraces are established, the moisture that accumulates behind them can help support vegetation, including trees that can both stabilise the ridges and produce fruit and nuts or fuel wood.

The terraces, which are particularly well adapted to the hilly agricultural regions of semiarid regions of India Africa, can markedly raise land productivity because they conserve both water and soil. The practice of collecting rain water where it falls is several hundred years old. In the desert areas all over the world, where there is a shortage, communities have been collecting rain water in open tanks, in open wells and from the roof tops of buildings for centuries.

With the coming of more so called advanced technology including sophisticated groundwater survey and exploration equipment, with the installation of hand pumps and piped water supply schemes, the importance of people's technologies and community solutions was devalued. There is thus a need to go back to learn the important lessons from the past on how communities have solved their drinking water problems by collecting rain water on a large scale.

Methods of Rainwater Harvesting in Cities: Broadly the rain water can be harvested by two methods:

1. Store the rainwater in containers above or above grounds or below grounds; and
2. Recharge into soil for withdrawal later by groundwater recharging basis.

Elements of typical water harvesting system: Any rain water harvesting will have four elements:

1. Catchment area;
2. Conduits;
3. Settlement Tank; and
4. Recharge facility or storage facility.

Rainwater harvesting techniques are simple, labourintensive and cost effective. By collecting water where it falls:

1. In underground tanks in rural primary schools.
2. In artificial ponds deepened to collect more water.
3. In unused and disused open wells where surface water is channelled for faster percolation into the ground, rainwater could be usefully harvested and at much lower costs.

Inter-state River Linkages

Inter-state river linkages are a topic of interest for India, especially for the southern States of the Indian Union where the rivers are largely non-perennial and the water shortages are alarming in the off-season and summer. There are similar contexts elsewhere in the world, in almost all continents. It is in India, the topic is hot and is being talked about intensely in the last few months. The present scenario in the country's water resources sector is alarmingly dismal. While floods are wreaking havoc in the Northeast and Eastern regions and drought is looming large in many parts of the Indian landmass, many States continue opposing the inter-basin transfer of waters from surplus to deficit basins.

Punjab has gone one step ahead by passing unilaterally a Bill terminating all previous agreements and accords on river waters thereby vitally affecting irrigation and drinking water supply in the neighbouring States. Punjab's action has triggered a whirlwind of protest and has raised a core issue of national importance, the solution to which would determine the future course of action on such issues. Conflict of interest is the normal state of affairs in a reality where river flows physically link upstream and downstream users and uses.

The issues are complex and linkages are many. But instead of sorting out the differences, of late, water endowed States have been resorting to legal gimmickry while playing to the political galleries, in their bid to prevent use of waters flowing through their territories by their water deficit neighbours. Kerala has passed a Bill in this regard recently and Karnataka had issued an ordinance on the Cauvery waters some time back.

In all these cases, reference has been made to the sovereign rights of the States as enshrined in the Constitution. If other water surplus States take the cue from these instances and act as arbitrarily, the consequences of such developments in terms of India's unity and integrity would be disastrous. Hence the demand has been gaining momentum in the last few years for the Central government to transfer the subject of "Water" from the "State List" to the "Union List" or "Concurrent List" to arrest further deterioration in water related issues.

In the Constitution, "Water, that is to say, water supplies, irrigation and canals, drainage and embankments, water storage and water power," is a matter contained in Entry 17, List II (State List). Hence the State legislatures have full powers to legislate under this provision on all water related matters including their regulation and development. However, this Entry is subject to the provision of Entry 56, List I (Union List) which authorises Parliament to enact laws for the regulation and development of interstate rivers and river valleys.

Regulation and development of interstate rivers and river valleys to the extent to which such regulation and development under the control of the Union is declared by Parliament by law to be expedient in public interest. In the past, the Centre had enacted over-riding laws using constitutional powers on many subjects such as industry and business, which are included in the State List. However, at present any constitutional amendment particularly in the emotive issue of water does not seem possible. Another view which has gained currency is that without any constitutional amendment, the Centre can deal with interstate rivers by empowering itself under the available provisions.

Against the above background, any change in the scheme of the Constitution has to be ruled out and instead the Centre has to pass laws to deal with interstate rivers more effectively as they contribute more than 85 per cent of the water resources of the country. The National Commission for Integrated Water Resources Development Plan had also recommended such an approach.

Consultative Mechanism: Apart from enacting laws to empower itself for the control and regulation of interstate rivers, there is also a need for the Centre to set up Centre-State consultative mechanisms for effectively sorting out of water related concerns. The purpose is to secure a proper tie between problems and their main causes on the one hand and technology and governance perspectives of problem- solving on the other.

Further, a National Water Authority (NWA) as an apex body to concern itself with the development and management of interstate rivers supported by a River Basin Organisation for each interstate river needs to be set up on similar lines as in Australia and France.

The River Boards Act has to be suitably amended to create these institutions. It must be remembered that the development of effective solutions to water problems depends more on governance. Hence it is essential for the Centre to empower itself to take over the interstate rivers for providing better regulation and management of the water resources available in the country and put a stop to the fissiparous tendencies recently demonstrated.

4

FOOD RESOURCES

Food may be sourced from plants, animals or other categories such as fungus or fermented products like alcohol. Although many human cultures sought food items through hunting and gathering, today most cultures use farming, ranching, and fishing, with hunting, foraging and other methods of a local nature included but playing a minor role.

Most traditions have a recognisable cuisine, a specific set of cooking traditions, preferences, and practices, the study of which is known as gastronomy. Many cultures have diversified their foods by means of preparation, cooking methods and manufacturing. This also includes a complex food trade which helps the cultures to economically survive by-way-of food, not just by consumption. Many cultures study the dietary analysis of food habits. While humans are omnivores, religion and social constructs such as morality often affect which foods they will consume. Food safety is also a concern with foodborne illness claiming many lives each year.

SOURCES OF FOOD

Almost all foods are of plant or animal origin, although there are some exceptions. Foods not coming from animal or plant sources include various edible fungi, including mushrooms. Fungi and ambient bacteria are used in the preparation of fermented and pickled foods such as leavened bread, wine, beer, cheese, pickles, and yogurt. Many cultures eat seaweed, which is a protist, or blue-green algae such as Spirulina. Additionally, salt is often eaten as a flavouring or preservative, and baking soda is used in food preparation. Both of these are inorganic substances, as is water, an important part of human diet.

Plants

Many plants or plant parts are eaten as food. There are around 2,000 plant species which are cultivated for food, and many have several distinct cultivars. Plant-based foods can be classified as with the nutrients necessary for the plant's initial growth. Because of this, seeds are often packed with energy, and are good sources of food for animals, including humans. In fact, the majority of all foods consumed by human beings are seeds. These include cereals, legumes, and nuts. Oilseeds are often pressed to produce rich oils, including sunflower, rape, and sesame.

Fruits are the ripened extensions of plants, including the seeds within. Fruits are made attractive to animals so that animals will eat the fruits and excrete the seeds over long distances. Fruits, therefore, make up a significant part of the diets of most cultures. Some fruits, such as pumpkin and eggplant, are eaten as vegetables. Vegetables are a second type of plant matter eaten as food. These include root vegetables, leaf vegetables, stem vegetables, and inflorescence vegetables. Many herbs and spices are highly-flavourful vegetables.

Animals

Animals can be used as food either directly, or indirectly

by the products they produce. Meat is an example of a direct product taken from an animal, which comes from either muscle systems or from organs. Food products produced by animals include milk produced by Mammals, which in many cultures is drunk or processed into dairy products such as cheese or butter. In addition birds and other animals lay eggs, which are often eaten, and bees produce honey, a popular sweetener in many cultures. Some cultures consume blood, some in the form of blood sausage, as a thickener for sauces, a cured salted form for times of food scarcity, and others use blood in stews such as civet.

Food Production

Food is traditionally obtained through farming, ranching, and fishing, with hunting, foraging and other methods of subsistence locally important. More recently, there has been a growing trend towards more sustainable agricultural practices. This approach, which is partly fuelled by consumer demand, encourages biodiversity, local self-reliance and organic farming methods. Major influences on food production are international organisations, national government policy, and war.

Preparation of Food

While some food can be eaten raw, many foods undergo some form of preparation for reasons of safety, palatability, or flavour. At the simplest level this may involve washing, cutting, trimming or adding other foods or ingredients, such as spices. It may also involve mixing, heating or cooling, pressure cooking, fermentation, or combination with other food. In a home, most food preparation takes place in a kitchen. Some preparation is done to enhance the taste or aesthetic appeal; other preparation may help to preserve the food; and others may be involved in cultural identity. A meal is made up of food which is prepared to be eaten at a specific time and place.

Slaughter and Butchering

The preparation of animal-based food will usually involve slaughter, evisceration, hanging, portioning and rendering. In developed countries, this is usually done outside the home in slaughterhouses which are used to process animals mass for meat production.

Many countries regulate their slaughterhouses by law. For example, the United States has established the Humane Slaughter Act of 1958, which requires that an animal be stunned before killing. This act, like those in many countries, exempts slaughter in accordance to religious law, such as kosher shechita and dhabiha halal.

Strict interpretations of kashrut require the animal to be fully aware when its carotid artery is cut. On the local level a butcher may commonly break down larger animal meat into smaller manageable cuts and pre-wrapped for commercial sale or wrapped to order in butcher paper. In addition fish and seafood may be fabricated into smaller cuts by a fish monger at the local level. However fish butchery may be done on board a fishing vessel and quick-frozen for preservation of quality.

World Food Situation

While 2004 and 2005 were promising in terms of renewed attention to reducing hunger, the world still faces the large and familiar problems of widespread hunger and malnutrition. In addition, it is becoming clearer that hidden hunger due to micronutrient deficiencies is widespread. At the same time, over-consumption and the chronic disease problems that accompany it are creeping into poor families. Progress toward reducing hunger has been slow in the past decade.

During the 1990s, developing countries reduced the number of undernourished people by only 9 million, representing a mere 1 per cent of the total of undernourished people. The progress in reducing hunger has been uneven across regions and countries. While China has been able to cut the number of undernourished since 1990, the rest of the developing world shows increased hunger.

The situation in Sub-Saharan Africa is particularly dire with the number of hungry people increasing by 20 per cent since 1990. Similarly the number of underweight children is estimated to have increased in most parts of Africa between 2000 and 2005, while there were noted decreases in all other developing-country regions. Micronutrient deficiencies, pose a vast global health problem. Vitamin A deficiency, iron deficiency anaemia, and zinc deficiency increase the probability of early death for children and women, impair IQ development in children, and lead to a large loss in quality of life, productivity, and economic growth in developing countries.

Iron deficiency anaemia affects 70 per cent of nonpregnant women in India and almost 50 per cent in Sub-Saharan Africa. Vitamin A deficiency affects the immune system of approximately 40 per cent of children under five years of age living in developing countries and leads to approximately 1 million child deaths every year. In some countries the impact is more severe; in India almost 60 per cent of preschool children suffer from vitamin A deficiency.

Food Crises and Disasters: In December 2004, countries bordering the Indian Ocean were swept by a tsunami that claimed the lives of more than 280,000 people and displaced 1.5 million people from their homes. Fisheries and coastal agriculture were severely damaged. In Sri Lanka, for example, 75 per cent of the fishery sector was affected, with many small fishers losing their lives and most losing their equipment and gear. In India, 300,000 fishers were made jobless. Furthermore, productive lands have been lost due to massive erosion and salinity infusion.

The Indian Ocean tsunami is one in a series of recent powerful natural disasters affecting developed and developing countries—from unexpected occurrences like earthquakes, floods, and cyclones to less immediate and evolving hazards like droughts and environmental degradation. The October 2005 South Asia Earthquake affecting Pakistan and India has resulted thus far in about 75,000 human casualties, and about $330 million loss in livestock, crops, and agricultural equipment.

The trend in the past two decades shows an increase in both the number of natural disasters and the number of people killed by them. In the past two years the number of deaths from natural disasters has been far above the past decade's average. Other disasters that lead to food crises are manmade. Large numbers of people throughout the developing world are still affected by armed conflicts and wars. Fortunately, data show that since the early 1990s, the number of armed conflicts has fallen by more than 40 per cent and the number of wars has fallen by 80 per cent. Nevertheless, there are continuing conflicts—such as in Darfur and Iraq—that remain a source of poverty and hunger for hundreds of thousands of people.

The 2004–05 food crisis in the Sahel—especially in Niger—demonstrates the limitations of the current global humanitarian response system. In fall 2004, following inadequate rains and a serious locust invasion, Niger experienced 75–80 per cent increases in grain prices and severe food shortages in some areas. NGOs working in Niger reported child malnutrition rates of two to six times normal levels. Yet early warnings and aid appeals from the government and the United Nations in 2004 and early 2005 went largely unheeded for far too long. A robust response to the first of these repeated calls for help would have meant the difference between a humanitarian catastrophe and a serious food problem.

World Food Production: World cereal production in 2004 exceeded 2 billion tons—a record level and an increase of approximately 9 per cent over the previous year. Most of the increase occurred in developed countries, in particular the United States and the European Union. Developing countries experienced a rise in cereal production of 3 per cent. The large increase in cereal production led to a build-up of world cereal stocks, the first increase of stocks in five years. But cereal production is expected to decline in 2005, indicating continued need for attention to the impacts of variability.

The global area planted with biotech Genetically Modified Organism (GMO) crops has steadily increased during the past nine years. By 2004, about 8.3 million farmers in 17 countries were growing such crops. Developing countries participated in this growth representing now more than 30 per cent of all area planted with biotech crops. Furthermore, the growth of acreage under biotech crops in developing countries exceeded that of developed countries for the first time during the past nine years.

Biotech products developed by public-sector agricultural research systems of developing countries are increasingly appearing in developing countries. The work of the Consultative Group on International Agriculture Research (CGIAR) centers on biosafety and biodiversity remains critical for sustained long-term agricultural growth. Simply noting the old and new developments of the world food situation gives only limited guidance for responding to opportunities and threats. We therefore have to look into the changing forces that drive the world food situation.

Drivers Affecting the World Food Situation

Today, the world food situation is subject to a number of new developments, such as new global and national policy initiatives, changes in the global agri food business, and rising energy prices.

Aid and Global Policy: Since the adoption of the Millennium Development Goals (MDGs) in September 2000, poverty has climbed to the top of the global development agenda, leading to a paradigm shift in development thinking and economic policies across the globe. The year 2005 saw a remarkable array of high-level initiatives for addressing poverty and hunger and a new focus on agriculture and rural areas, where most hunger persists.

The United Nations and its Millennium Project issued several new reports and guidelines related to the MDGs in 2005. Among them was a report by the Task Force on Hunger, *Halving Hunger: It Can Be Done*, which presented an agenda of the highest priorities for action to be considered for cutting hunger in half. These priorities consisted of policy reforms; the creation of an enabling environment; and measures to increase the agricultural productivity of food-insecure farmers, improve nutrition for the chronically hungry and vulnerable, reduce the vulnerability of the acutely hungry through productive safety nets, and increase incomes and make markets work for the poor.

In March 2005, the Commission for Africa, chaired by Britis Prime Minister Blair, released its report *Our Common Interest*. The Commission made strong recommendations in a number of areas, notably governance and capacity building as well as peace and security. It called for doubling aid to Sub-Saharan Africa—that is, increasing aid by an additional US$25 billion per year by 2010. The Commission also recommended a review of progress in 2010, with a possible second stage involving a further US$25 billion a year to be issued by 2015. For poor countries in Sub-Saharan Africa, it called for 100 per cent debt cancellation.

Africa also featured prominently at the G8 Summit, held in July 2005 in Gleneagles, Scotland. Along the lines of the Commission for Africa recommendations, the G8 leaders agreed to:

1. Double aid for Africa by US$25 billion a year by 2010; and
2. Cancel 100 per cent of the multilateral debts of the highly indebted poor countries (HIPC) and adopt a special package of debt cancellation for Nigeria worth around US$17 billion.

If actually implemented, these decisions on aid for Africa constitute a major accomplishment. There are concerns, however, about whether all these commitments and pledges will translate into actual new disbursements rather than money shuffled from other aid uses. Changing an unfortunate trend of past years, the World Bank's investment for agriculture and rural development increased in 2004/05.

Global trade policy: Aid alone will not facilitate the achieving of the MDGs. Creating improved trade opportunities for the developing countries are a key element of the agenda. The Doha Development Round of the WTO is yet to show results for farmers in developing countries by reducing domestic support in OECD countries and increasing market access. But the negotiators thus far have missed every deadline to reach agreement on reducing agricultural subsidies and other protectionist measures that hurt small-scale farmers in developing countries. It is crucial to strengthen a rules-based system of fair trade through a three-pronged approach:

1. Industrialised countries should improve access to their markets;
2. Developing countries should do their part, as well, to reduce trade restrictions; and
3. Wealthy countries should provide more assistance to connect poor farmers to markets.

Food safety concerns are and will be even more important for livestock product trade which offers so much opportunity for small farmers. But today large parts of international trade are restricted because of animal diseases and the threat of avian flu looms large. The CGIAR's work on animal diseases has an even more important role to play in addressing these concerns.

National and regional policy initiatives: International initiatives can only set the context for appropriate action at the country level, where policies are formulated, budgets are set, and programs are established. In 2005, some of the world's largest developing countries made significant policy improvements, driven largely by domestic political pressure. China is increasingly paying attention to issues of equity, not just growth.

In September 2005 the Politburo, China's top policymaking body, pledged to spread the benefits of economic growth more fairly among all levels of Chinese society, and in particular to close the gap between cities and the countryside. India's 2005–06 budget has allocated substantially more resources for agriculture and rural development, including water resource management and programs for education, health, and employment for poverty reduction. Brazil has substantially increased resources to its Fome Zero (Zero Hunger) programme from Rs. 5.7 billion in 2003 to Rs. 12.2 billion for 2005. Because of the time requirements of the budgetary process, it is only now in 2005, two years after the initiative was announced, that large-scale effects can be expected.

African governments are beginning to effectively work together through a revitalised African Union and its flagship programme, the New Partnership for Africa's Development (NEPAD). However, while the stated goal of increasing the budget allocation to agriculture is often referred to, it is not clear how progress will be monitored or how donors can ensure that increased aid to Africa is used for the intended MDG-related purposes. The established peer review mechanism under NEPAD may help with that.

A changing corporate food system: Not only the public sector, but also the private sector is now an important player in the changing world food situation, and the related innovation and research system. The corporate food system is increasing its market share in developing countries, where urbanisation is creating new retail food consumers. Indeed, the fastest-growing element in the food chain in developing countries is the supermarket sector with growth rates in sales exceeding 20 per cent per annum in some countries.

In addition, food manufacturing and processing are on the rise, as urban consumers demand more processed foods. Industry is increasingly looking at the millions of small farmers and poor consumers as customers. In some regions, such as East Asia, small farmers are getting a foothold in this changing food chain, but the smaller they are, the higher their unit transaction costs to participate in the food chain. Companies along the food chain are becoming locally increasingly concentrated, highly coordinated along the food chain, and more global in their operations.

The world's farmers, of which 85 per cent have less than 2 hectares, must link in new and efficient ways to input, processing, and retail industries to capture a fair share in the value chain. The concentration rate is a matter of concern to many. However, at a global scale, even the allegedly highly concentrated retail sector's top five companies do not yet capture more than 14 per cent of the market. We shall continue to monitor these developments for their opportunities and risks.

At the same time, rising consumer incomes and changing lifestyles are creating bigger markets for high-value agricultural products like fruits, vegetables, fish, and meat. The growing markets for these products present an opportunity for developing-country farmers to diversify their production out of staple grains and raise their incomes. Annual growth rates on the order of 8 to 10 per cent in high-value agricultural products are a promising development, as the production, processing, and marketing of these products create a lot of needed employment in rural areas.

Food prices and energy: We have become used to a secular long-term decline in world staple food prices over past decades. However, this trend, which has been fostered by technological progress, especially in plant breeding, is not visible in the past decade.

A major new factor with short- and long-term implications for agriculture and thus poverty and hunger is the price of energy. As energy prices rise, the costs of farmers' inputs, processing, and transportation go up. This could result in an increase in food price, with varying effects for different people. So far we do not see such effects, as food prices do not yet correlate significantly with energy prices. However, the situation may become fundamentally different because it is becoming increasingly efficient to produce energy from agricultural products. At current oil prices it is cost-effective to produce bio-fuel. Competition for land and other resources will rise between production for food versus for fuel. Many developing countries are currently looking at that option and considering or already implementing expanded bio-fuel production and processing capacities. New challenges for accelerated innovation in agriculture and in the energy sector (including for saving energy) arise from that.

Achieving the MDGs

The coming years must see strong initiatives to actually implement the actions for reaching the MDGs. International Food Policy Research Institute (IFPRI) has developed a scenario to explore the agricultural development and investment needs required for developing countries to halve the proportion of childhood malnutrition. This MDG-compatible scenario is based on its International Model for Policy Analysis of Agricultural Commodities and Trade (IMPACT). It envisions reducing the proportion of child malnutrition by half for nearly all of the developing countries through investments in five key drivers of food and nutrition security: rural road construction, education, clean water provision, agricultural research, and irrigation.

IFPRI-IMPACT projects that with business as usual the proportion of children who are malnourished would drop to 24 per cent in 2015, whereas under the MDG scenario it would drop to 17 per cent. Sub-Saharan Africa, South Asia, and parts of Southeast Asia are expected to struggle to halve the proportion of child malnutrition whereas China and Latin America are well on their way toward significant continued reduction in child malnutrition. IFPRI-IMPACT projects that the business as usual scenario will cost US$21.5 billion per annum during 1995-2015 for all developing countries. The MDG scenario projects US$29.6 billion per annum over the same period. The investments are needed in the five key drivers—rural road construction, education, clean water provision, agricultural research, and irrigation. Essentially, $8 billion more per annum could bring us close to halving the proportion of child malnutrition by 2015.

5

STATE OF GLOBAL AGRICULTURE

The history of agriculture is a central element of human history, as agricultural progress has been a crucial factor in worldwide socio-economic change. Wealth-building and militaristic specialisations rarely seen in hunter-gatherer cultures are commonplace in agricultural and agro-industrial societies—when farmers became capable of producing food beyond the needs of their own families, others in the tribe/village/City-state/nation/empire were freed to devote themselves to projects other than food acquisition.

As of 2006, an estimated 36 per cent of the world's workers are employed in agriculture, making it by far the most common occupation. However, the relative significance of farming has dropped steadily since the beginning of industrialisation, and in 2006-for the first time in history-the services sector overtook agriculture as the economic sector employing the most people worldwide. Also, agricultural production accounts for less than five per cent of the gross world product.

Agriculture played a key role in the development of human civilisation—it is widely believed that the domestication of plants and animals allowed humans to settle and give up their previous hunter-gatherer lifestyle during the Neolithic Revolution. Until the Industrial

Revolution, the vast majority of the human population laboured in agriculture. Development of agricultural techniques has steadily increased agricultural productivity, and the widespread diffusion of these techniques during a time period is often called an agricultural revolution. A remarkable shift in agricultural practices has occurred over the past century in response to new technologies.

In particular, the Haber-Bosch method for synthesizing ammonium nitrate made the traditional practice of recycling nutrients with crop rotation and animal manure less necessary. Synthetic nitrogen, along with mined rock phosphate, pesticides and mechanisation, have greatly increased crop yields in the early 20th century. Increased supply of grains has led to cheaper livestock as well. Further, global yield increases were experienced later in the 20th century when high-yield varieties of common staple grains such as rice, wheat, and corn were introduced as a part of the Green Revolution. The Green Revolution exported the technologies of the developed world out to the developing world.

AGRICULTURAL OUTPUT

Many governments have subsidised agriculture to ensure an adequate food supply. These agricultural subsidies are often linked to the production of certain commodities such as wheat, corn, rice, soybeans, and milk. These subsidies, especially when done by developed countries have been noted as protectionist, inefficient, and environmentally damaging. In the past century agriculture has been characterised by enhanced productivity, the use of synthetic fertilisers and pesticides, selective breeding, mechanisation, water contamination, and farm subsidies.

Proponents of organic farming such as Sir Albert Howard argued in the early 1900s that the overuse of pesticides and synthetic fertilisers damages the long-term fertility of the soil. While this feeling lay dormant for decades, as environmental awareness has increased recently there has been a movement towards sustainable agriculture by some farmers, consumers, and policymakers. In recent years there has been a backlash against perceived external environmental effects of mainstream agriculture, particularly regarding water pollution, resulting in the organic movement.

One of the major forces behind this movement has been the European Union, which first certified organic food in 1991 and began reform of its Common Agricultural Policy (CAP) in 2005 to phase out commodity-linked farm subsidies, also known as decoupling. The growth of organic farming has renewed research in alternative technologies such as integrated pest management and selective breeding. Recent mainstream technological developments include genetically modified food. As of late 2007, increased farming for use in biofuels has pushed up the price of grain used to feed poultry and dairy cows and other cattle, causing higher prices of wheat, soybean, and maize over the year.

Agricultural Practices

Agricultural practices lie on a spectrum dependent upon the intensity and technology of the methods. At the one end lies the subsistence farmer who farms a small area with limited inputs and produces only enough food to meet the needs of his or her family. At the other end lies intensive agriculture which includes traditional labour intensive farming, and modern agriculture which includes industrial agriculture, organic farming and sustainable farming.

Industrial agriculture involves large fields and/or numbers of animals, high resource inputs, and a high level of mechanisation. These operations achieve economies of scale and require large amounts of capital in the form of land and machinery. The twentieth century saw changes in agricultural practice, particularly in agricultural chemistry and in mechanisation. Agricultural chemistry includes the application of chemical fertiliser, chemical insecticides, and chemical fungicides, analysis of soil makeup and nutritional needs of farm animals.

Mechanisation has increased farm efficiency and productivity in most regions of the world, due especially to the tractor and various "gins" (short for "engine") such as the cotton gin, semi-automatic balers and threshers and, above all, the combine. According to the National Academy of Engineering in the United States, agricultural mechanisation is one of the 20 greatest engineering achievements of the 20th century. Early in the century, it took one American farmer to produce food for 2.5 people. By 1999, due to advances in agricultural technology, a single farmer could feed over 130 people. Other recent changes in agriculture include hydroponics, plant breeding, hybridisation, gene manipulation, better management of soil nutrients, and improved weed control.

Genetic engineering has yielded crops which have capabilities beyond those of naturally occurring plants, such as higher yields and disease resistance. Modified seeds germinate faster, and thus can be grown on an accelerated schedule. Genetic engineering of plants has proven controversial, particularly in the case of herbicide-resistant plants. Genetic engineers at companies such as Monsanto are working to develop plants for irrigation, drainage, conservation and sanitary engineering, particularly important in normally arid areas which rely upon constant irrigation, and on large scale farms.

The processing, packing and marketing of agricultural products are closely related activities also influenced by science. Methods of quick-freezing and dehydration have increased the markets for many farm products. Animals, including horses, mules, oxen, camels, llamas, alpacas, and dogs, are often used to help cultivate fields, harvest crops, wrangle other animals, and transport farm products to buyers. Animal husbandry not only refers to the breeding and raising of animals for meat or to harvest animal products on a continual basis, but also to the breeding and care of species for work and companionship.

Airplanes, helicopters, trucks, tractors, and combines are used in Western agriculture for seeding, spraying operations for insect and disease control, harvesting, aerial topdressing and transporting perishable products. Radio and television disseminate vital weather reports and other information such as market reports that concern farmers. Computers have become an essential tool for farm management. In recent years, some aspects of intensive industrial agriculture have been the subject of increasing debate.

The widening sphere of influence held by large seed and chemical companies, meat packers and food processors has been a source of concern both within the farming community and for the general public. Another issue is the type of feed given to some animals that can cause bovine spongiform encephalopathy in cattle. There has also been concern over the effect of intensive agriculture on the environment.

The patent protection given to companies that develop new types of seed using genetic engineering has allowed seed to be licensed to farmers in much the same way that computer software is licensed to users. This has changed the balance of power in favour of the seed companies, allowing them to dictate terms and conditions previously unheard of. The Indian activist and scientist Vandana Shiva argues that these companies are guilty of biopiracy.

Soil conservation and nutrient management have been important concerns since the 1950s, with the most advanced farmers taking a stewardship role with the land they use. However, increasing contamination of waterways and wetlands by nutrients like nitrogen and phosphorus are concerns that can only be addressed by "enlightenment" of farmers and/or far stricter law enforcement in many countries. Increasing consumer awareness of agricultural issues has led to the rise of community-supported agriculture, local food movement, "Slow Food", and commercial organic farming.

Crop Plants

Domestication of plants has, over the centuries increased yield, improved disease resistance and drought tolerance, eased harvest and improved the taste and nutritional value of crop plants. Careful selection and breeding have had enormous effects on the characteristics of crop plants. Plant breeders use greenhouses and other techniques to get as many as three generations of plants per year towards the continued effort of improvement. Plant selection and breeding in the 1920s and 1930s improved pasture in New Zealand. Extensive X-ray an ultraviolet induced mutagenesis efforts during the 1950s produced the modern commercial varieties of grains such as wheat, corn and barley.

After mechanical tomato-harvesters were developed in the early 1960s, agricultural scientists bred tomatoes that were more resistant to mechanical handling. These varieties have been criticised as being harder and having poor texture. More recently, genetic engineering has begun to be employed in large parts of the world to speed up the selection and breeding process. One widely used modification is a herbicide resistance gene that allows plants to tolerate exposure to glyphosate, a non-systemic chemical used to control weeds in a crop such as oilseed rape.

Normally, expensive systemic herbicides would have to be applied to kill the weeds without harming the crop. Relatively cheap and safe glyphosate may be applied to the modified crops, efficiently killing weeds without harming the resistant crop. Another modification causes the plant to produce a toxin to reduce damage from insects.

FARMING PRACTICES OF LIVESTOCK

The farming practices of livestock vary dramatically world-wide and between different types of animals. Livestock are generally kept in an enclosure, are fed by human-provided food and are intentionally bred, but some livestock are not enclosed, or are fed by access to natural foods, or are allowed to breed freely, or all three. Approximately 68 per cent of all agricultural land is used in the production of livestock as permanent pastures. Livestock' are defined, in part, by their end purpose as the production of food or fibre, or labour. The economic value of livestock includes:

Meat: The production of a useful form of dietary protein and energy.

Dairy products: Mammalian livestock can be used as a source of milk, which can in turn easily be processed into other dairy products such as yogurt, cheese, butter, ice cream, kefir, and kumis. Using livestock for this purpose can often yield several times the food energy of slaughtering the animal outright.

Fibre: Livestock produce a range of fibre/textiles. For example, sheep and goats produce wool and mohair; cows, deer, and sheep can make leather; and bones, hooves and horns of livestock can be used.

Fertiliser: Manure can be spread on fields to increase crop yields. This is an important reason why historically, plant and animal domestication have been intimately linked. Manure is also used to make plaster for walls and floors and can be used as a fuel for fires. The blood and bone of animals are also used as fertiliser.

Labour: Animals such as horses, donkey, camels and yaks can be used for mechanical energy. Prior to steam power livestock were the only available source of non-human labour. They are still used for this purpose in many places of the world, including ploughing fields, transporting goods, and military functions.

Land management: The grazing of livestock is sometimes used as a way to control weeds and undergrowth. For example, in areas prone to wild fires, goats and sheep are set to graze on dry scrub which removes combustible material and reduces the risk of fires.

During the history of animal husbandry many secondary products have arisen in an attempt to increase carcass utilisation and reduce waste. For example, animal offal and non-edible parts may be transformed into products such as pet food and fertiliser. In the past such waste products were sometimes also fed to livestock as well. However, intra-species recycling poses a disease risk, threatening animal and even human health.

Farming practices vary dramatically world-wide and between types of animals. Livestock are generally kept in an enclosure, are fed by human-provided food and are intentionally bred, but some livestock are not enclosed, or are fed by access to natural foods, or are allowed to breed freely, or any combination thereof. Livestock raising historically was part of a nomadic or pastoral form of material culture. The herding of camels and reindeer in some parts of the world remain unassociated with sedentary agriculture.

The transhumance form of herding in the Sierra Nevada Mountains of California still continues as cattle, sheep or goats are moved from winter pasture in lower lying valleys to spring pasture and summer pasture in the foothills and alpine regions as the seasons progress. Cattle were raised on the open range in the Western United States and Canada, as well as on the Pampas of Argentina and other prairie and steppe regions of the world. The enclosure of livestock in pastures and barns is a relatively new development in the history of agriculture.

When cattle are enclosed, the type of 'enclosure' may vary from a small crate or to a large fenced pasture. The type of feed may vary from natural growing grass, to highly sophisticated processed feed. Animals are usually intentionally bred through artificial insemination or through supervised mating. Indoor production systems are generally used only for pigs and poultry, as well as for veal cattle. Indoor animals are generally farmed intensively, as large space requirements would make indoor farming unprofitable and impossible.

However, indoor farming systems are controversial due to: the waste they produce, odour problems, the potential for groundwater contamination and animal welfare concerns. Other livestock are farmed outside, although the size of enclosure and level of supervision may vary. In large open ranges animals may be only occasionally collected in "round-ups" or "musters". Herding dogs such as sheep dogs and cattle dogs may be used for mustering as are cowboys, musterers and jackaroos on horseback or in helicopters. Since the advent of barbed wire and electric fence technology, fencing pastures has become much more feasible and pasture management simplified.

Rotation of pasturage is a modern technique for improving nutrition and health while avoiding environmental damage to the land. In some cases very large numbers of animals may be kept in indoor or outdoor feeding operations, where the animals' feed is processed, offsite or onsite, and stored onsite then fed to the animals.

Livestock—especially cattle—may be branded to indicate ownership, but in modern farming identification is more likely to be indicated by means of ear tags than burning. This is not only more humane, but also has other advantages such as reducing the likelihood of infection and damage to the livestock.

Sheep are also frequently marked by means of ear tags. As fears of mad cow disease and other epidemic illnesses mount, the use of microchip identification to monitor and trace animals in the food production system is increasingly common, and sometimes required by governmental regulations. Modern farming techniques seek to minimise human involvement, increase yield, and improve animal health.

Economics, quality and consumer safety all play a role in how animals are raised. Drug use and feed supplements may be regulated, or prohibited, to ensure yield is not increased at the expense of consumer health, safety or animal welfare. Practices vary around the world, for example growth hormone use is permitted in the United States but not in the European Union or in countries selling meat/produce in the EU such as Australia and New Zealand.

Environmental Problems

Agriculture may often cause environmental problems because it changes natural environments and produces harmful by-products. Some of the negative effects are:

- Loss of biodiversity.
- Surplus of nitrogen and phosphorus in rivers and lakes.
- Detrimental effects of herbicides, fungicides, insecticides, and other biocides.
- Conversion of natural ecosystems of all types into arable land.
- Consolidation of diverse biomass into a few species.
- Soil erosion.
- Depletion of minerals in the soil.
- Particulate matter, including ammonia and ammonium off-gassing from animal waste contributing to air pollution.
- Weed Science—feral plants and animals.
- Odour from agricultural waste.
- Soil salination.

According to the United Nations, the livestock sector emerges as one of the top two or three most significant contributors to our most serious environmental problems, at every scale from local to global. Livestock production occupies 70 per cent of all land used for agriculture, or 30 per cent of the land surface of the planet.

It is one of the largest sources of greenhouse gases—responsible for 18 per cent of the world's greenhouse gas emissions as measured in CO_2 equivalents. By comparison, all transportation emits 13.5 per cent of the

CO_2. It produces 65 per cent of human-related nitrous oxide and 37 per cent of all human-induced methane. It also generates 64 per cent of the ammonia, which contributes to acid rain and acidification of ecosystems.

Genetic Erosion in Crops

Genetic erosion in crops and livestock biodiversity is propelled by several major factors such as variety replacement, land clearing, over exploitation of species, population pressure, environmental degradation, overgrazing, policy and changing agricultural systems. The main factor, however, is the replacement of local varieties of domestic plants and animals by high yielding or exotic varieties or species. A large number of varieties can also often be dramatically reduced when commercial varieties are introduced into traditional farming systems.

Many researchers believe that the main problem related to agro-ecosystem management is the general tendency towards genetic and ecological uniformity imposed by the development of modern agriculture.

In agriculture and animal husbandry, the green revolution popularised the use of conventional hybridisation to increase yield many folds by creating "high-yielding varieties". Often the handful of breeds of plants and animals hybridised originated in developed countries and were further hybridised with local varieties in the rest of the developing world to create high yield strains resistant to local climate and diseases.

Hybridisation of local breeds to improve performance may lead to the loss of the local breed over time and consequently the loss of the genetic material that adapted that breed specifically to the local conditions. When viewed across the world as a whole, the consequent loss in genetic diversity and biodiversity could be placing the food supply in jeopardy, as a highly specialised breed may not contain sufficient genetic material to adapt to new diseases or environments even with an intensive breeding programme.

A Genetically Modified Organism (GMO) is an organism whose genetic material has been altered using the genetic engineering techniques generally known as recombinant DNA technology. Genetic Engineering today has become another serious and alarming cause of genetic pollution because artificially created and genetically engineered plants and animals in laboratories, which could never have evolved in nature even with conventional hybridisation, can live and breed on their own and what is even more alarming interbreed with naturally evolved wild varieties.

Genetically Modified (GM) crops today have become a common source for genetic pollution, not only of wild varieties but also of other domesticated varieties derived from relatively natural hybridisation. Since the 1940s, agriculture has dramatically increased its productivity, due largely to the use of petrochemical derived pesticides, fertilisers, and increased mechanisation. This has allowed world population to grow more than double over the last 50 years. Every energy unit delivered in food grown using modern techniques requires over ten energy units to produce and deliver.

The vast majority of this energy input comes from fossil fuel sources. Because of modern agriculture's current heavy reliance on petrochemicals and mechanisation, there are warnings that the ever decreasing supply of oil will inflict major damage on the modern industrial agriculture system, and could cause large food shortages. Oil shortages are one of several factors making organic agriculture and other sustainable farming methods necessary. This conversion is now occurring, but the reconditioning of soil to restore nutrients lost during the use of monoculture agriculture techniques made possible by petroleum-based technology will take time.

Some farmers using modern organic-farming methods have reported yields as high as those available from conventional farming. Farmers have also begun raising crops such as corn for non-food use in an effort to help mitigate peak oil. This has led to a 60 per cent rise in wheat prices recently, and has been indicated as a possible precursor to "serious social unrest in developing countries." Such situations would be exacerbated in the event of future rises in food and fuel costs, factors which have already impacted the ability of charitable donors to send food aid to starving populations.

Agriculture Safety and Health

Agriculture ranks among the most hazardous industries. Farmers are at high risk for fatal and nonfatal injuries, work-related lung diseases, noise-induced hearing loss, skin diseases, and certain cancers associated with chemical use and prolonged sun exposure. Farming is one of the few industries in which the families are also at risk for injuries, illness, and death.

In an average year, 516 workers die doing farm work in the U.S. Of these deaths, 101 are caused by tractor overturns. Every day, about 243 agricultural workers suffer lost-work-time injuries, and about 5 per cent of these result in permanent impairment.

Young Agricultural Workers

Agriculture is the most dangerous industry for young workers, accounting for 42 per cent of all work-related fatalities of young workers in the U.S. between 1992 and

2000. Unlike other industries, half the young victims in agriculture were under age 15. For young agricultural workers aged 15–17, the risk of fatal injury is four times the risk for young workers in other workplaces. Agricultural work exposes young workers to safety hazards such as machinery, confined spaces, work at elevations, and work around livestock.

An estimated 1.26 million children and adolescents under 20 years of age resided on farms in 2004, with about 699,000 of these youth performing work on the farms. In addition to the youth who live on farms, an additional 337,000 children and adolescents were hired to work on U.S. farms in 2004. On average, 103 children are killed annually on farms. Approximately 40 per cent of these deaths were work-related. In 2004, an estimated 27,600 children and adolescents were injured on farms; 8,100 of these injuries were due to farm work.

PESTICIDES

The widespread use and disposal of pesticides by farmers, large plantations and the general public causes environmental contamination. It is estimated that the 68 per cent of such contamination is a result of agricultural uses, followed by industrial and commercial uses and home and garden application following release into the environment, pesticides may give rise to different consequences. Pesticides which are sprayed can become airborne and may eventually end up in soil or water. Pesticides applied directly to the soil may be washed off the soil into water or may percolate through the soil to lower soil layers and groundwater.

The application of pesticides directly to bodies of water for weed control or indirectly as a result of run-off from soil or other routes, may lead not only to the build-up pesticides in water, but may also cause pollution to the air through evaporation. As these pesticides may be broken down, or degraded, by the action of sunlight, water, or other chemicals or microorganisms. This degradation process usually leads to the formation of less harmful residue but in some instances can produce more toxic products.

The second possibility is that the pesticides become resistant to degradation by any means and thus remain in unchanged in the environment for long periods of time. The ones that are most rapidly broken down have the shortest time to move or will otherwise have adverse effects on humans or other organisms. The ones which last the longest, the so-called persistent pesticides, can build up in the environment, leading to greater potential for adverse effects to occur.

Properties

In addition to resistance to degradation, there are a number of other properties of pesticides which determine their behaviour. Among these properties, the most important is the volatility—the rate of evaporation—of the pesticide. The ones that are most volatile have the greatest potential to go into the atmosphere and travel far if aided by the wind. Another important property is solubility in water. If a pesticide is very soluble in water, it seeps more easily through the soil and is a potential groundwater contaminant.

In addition, the water-soluble pesticide is more likely to stay in the water where it can have adverse effects on fish and other organisms. If the pesticide is insoluble in water, it usually tends to stick to the soil and will settle to the bottoms of bodies of water, making it less available to organisms.

Table 5.1 lists some of the more commonly used pesticides with an estimate of their persistence in soil. A standard approach to measure this parameter is the time taken for 75% to 100% of the material to disappear from the site of application. Using this classification, the grouping of the major classes of pesticides can be divided into non-persistent pesticides in the range of one to 12 weeks, moderately persistent pesticide and persistent pesticides.

Table 5.1: Pesticide Persistence in Soils

Low persistence	*Moderate persistence*	*High persistence*
Captan	Atrazine	Picloram
Dalapon	Carbaryl	Bromacil
Dicamba	Carbofuran	Chlordane
Malathion	Diazinon	Lindane
2, 4-D	Glyphosate	Paraquat
Dalapon	Carbaryl	Bromacil

If more persistent pesticides such as organochlorines are released into the environment, the potential hazard created to both land and acquatic ecosystems would be higher because the organisms would be exposed for long periods of time after a single application. A non-persistent pesticide, on the other hand, would not have a chance to move far in the environment or exert its effects for a longer period of time.

However, if it is also very soluble in water and the condition are right, it can move rapidly through certain soils. As it moves away from the surface, it moves away from the agents such as sunlight and bacteria which are degrading it. As it gets deeper into the soil, it degrades more slowly and thus has a chance to get into groundwater.

The dawnward movement of non-persistent pesticides is not an unlikely scenario and several pesticides with short half-lives, such as aldicarb, have been widely found in groundwater. In contrast, very persistent pesticides may have other properties which limit their potential for movement throughout the environment. many of the chlorinated hydrocarbon pesticides are very persistent and slow to break down but are also water-soluble and tend not to move down through the soil into groundwater.

They can, however, pose problems in other ways since they remain on the surface for a long time where they may be subject to run-off and possible evaporation. Even if they are not very volatile, the tremendously long time they persist can lead, over time, to measurable concentrations moving through the atmosphere and accumulating in remote areas.

Pesticides Effects on Living Organisms

Living organisms may also play a significant role in the movement of pesticides in the environment. This is particularly important for pesticides which can accumulate in living creatures.

An example to illustrate this accumulation is the uptake of a very water-soluble pesticide by a creature living in water. Since this pesticide is stored in the organism, the pesticide accumulates and levels increase over time. If this organism is eaten by a higher organism which can also store this pesticide, levels can reach higher values in the higher organism than is present in the water in which it lives.

Bioaccumulation is a phenomenon by which the concentration of a chemical is increased in successive steps of a food chain, with the danger that some species, especially the ones highest in the series, will be injured by the compound. A classical example of this injury is the population decline of fish-eating birds cause by reproductive failure as we as the thinning of the eggshells of these birds caused by organochlorine insecticides.

Similarly, it should be remembered that humans are at the top of the food chain and may be exposed to these high levels when we eat animal-based foods which have bioaccumulated pesticides and other organic chemicals. Apart from fish, domestic animals too can be accumulators of pesticides. Therefore, caremust be taken in the use of pesticides in agricultural activities. The release of pesticides into the environment may be followed by a complex series of events which can transport the pesticide through air or water, into the ground or even living organism.

The most important route of distribution and the extent of distribution are different for each pesticide. It will depend on the formulation of the pesticide and how and when it is released. Despite this complexity, it is possible to identify situations that can pose concern and to try to minimise them. However, there are significant gaps in the knowledge of pesticide movement in the environment. So it is best to minimise unnecessary release of pesticide into the environment. The fewer pesticides that are unnecessarily released, the safer our environment will be.

GOALS OF SUSTAINABLE AGRICULTURE

Sustainable agriculture integrates three main goals: environmental stewardship, farm profitability, and prosperous farming communities. These goals have been defined by a variety of disciplines and may be looked at from the vantage point of the farmer or the consumer.

Sustainable agriculture refers to the ability of a farm to produce food indefinitely, without causing irreversible damage to ecosystem health. Two key issues are biophysical and socio-economic. The physical aspects of sustainability are partly understood. Practices that can cause long-term damage to soil include excessive tillage and irrigation without adequate drainage. Long-term experiments provide some of the best data on how various practices affect soil properties essential to sustainability.

While air and sunlight are generally available in most geographic locations, crops also depend on soil nutrients and the availability of water. When farmers grow and harvest crops, they remove some of these nutrients from the soil. Without replenishment, the land would suffer from nutrient depletion and be unusable for further farming. Sustainable agriculture depends on replenishing the soil while minimising the use of non-renewable resources, such as natural gas, or mineral ores. Possible sources of nitrogen that would, in principle, be available indefinitely, include:

1. Recycling crop waste and livestock or human manure,
2. Growing legume crops and forages such as, peanuts, or alfalfa that form symbioses with nitrogen-fixing bacteria called rhizobia,
3. Industrial production of nitrogen by the Haber Process uses hydrogen, which is currently derived from natural gas, but could instead be made by electrolysis of water using electricity, or
4. Genetically engineering (non-legume) crops to form nitrogen-fixing symbioses or fix nitrogen without microbial symbionts.

The last option was proposed in the 1970s, but would be well beyond the capability of current technology, even if various concerns about biotechnology were addressed.

Sustainable options for replacing other nutrient inputs are more limited. In some areas, sufficient rainfall is available for crop growth, but many other areas require irrigation. For irrigation systems to be sustainable they must be managed properly and not use more water from their source than is naturally replenished, otherwise the water source becomes, in effect, a non-renewable resource.

Improvements in water well drilling technology and the development of submersible pumps have made it possible for large crops to be regularly grown where reliance on rainfall alone previously made this level of success unpredictable. However, this progress has come at a price, in that in many areas where this has occurred, such as the Ogallala Aquifer, the water is being used at a greater rate than its rate of recharge. Socioeconomic aspects of sustainability are also partly understood.

Economic Context

Given the finite supply of natural resources, agriculture that is inefficient may eventually exhaust the available resources or the ability to afford and acquire them. It may also generate negative externality, such as pollution as well as financial and production costs.

Agriculture that relies mainly on inputs that are extracted from the earth's crust or produced by society, contributes to the depletion and degradation of the environment. Despite this continuing practice, unsustainable agriculture continues because it is financially more cost-effective than sustainable agriculture in the short-term.

In an economic context, the need for the farm to generate revenue depends on the extent to which it is market oriented and on government subsidy. The way that crops are sold must be accounted for in the sustainability equation. Fresh food sold from a farm stand requires little additional energy, aside from that necessary for cultivation, harvest, and transportation. Food sold at a remote location, whether at a farmers' market or the supermarket, incurs a different set of energy cost for materials, labour, and transport.

To be sold at a remote location requires a complex economic system in which the farm producers form the first link in a chain of processors and handlers to the consumers. This practice allows greater revenue because of efficient transport of a large number of items, but because it produces externalities and relies on the use of non-renewable resources, shipping, processing, and handling, it is not sustainable. Moreover, such a system is considered vulnerable to fluctuations, such as strikes, oil prices, and global economic conditions including labour, interest rates, futures markets, and farm product prices.

In Third World agriculture, much of what is known about the social components of sustainability comes from anthropologist Robert Netting's work. In *Smallholders, Householders: Farm Families and the Ecology of Intensive, Sustainable Agriculture,* he defines an important cross-cultural pattern of high-labour, high-production cultivation exemplified East Asian paddy rice cultivators, African cultivators such as the Kofyar, alpine peasants, and Mesoamerican farmers of raised fields.

One key to socio-economic sustainability in such systems is that these farmers systems provide for much of their own subsistence and also participate in the market. From a system's view, the gain and loss factors for sustainability can be listed. The most important factors for an individual site are sun, air, soil and water as rainfall. These are naturally present in the system as part of the larger planetary processes and incur no costs. Of the four, soil quality and quantity are most amenable to human intervention through time and labour.

Natural growth and outputs are also subject to human intervention. What grows and how and where it is grown are a matter of choice. Two of the many possible practices of sustainable agriculture are crop rotation and soil amendment, both designed to ensure that crops being cultivated can obtain the necessary nutrients for healthy growth.

Monoculture

Monoculture, a method of growing only one crop at a time in a given field, is a very widespread practice, but there are questions about its sustainability, especially if the same crop is grown every year. Growing a mixture of crops sometimes reduces disease or pest problems but polyculture has rarely, if ever, been compared to the more widespread practice of growing different crops in successive years crop rotation with the same overall crop diversity.

Some pesticides, though sometimes useful in the short term, can harm the soil food web, a complex ecology of micro-organisms in soil that helps sustain the plant from the roots down. Experiments comparing plants grown in soil compared to plants grown through hydroponics have shown a thirty-three per cent higher growth rate when there are beneficial soil microorganisms available. Certain pesticides synthesised by chemical companies can impart a sometimes fatal toxicity to humans, livestock and insect pollinators, such as bees and butterflies, which may be necessary for plant success.

Without insect pollinators, farm labour must be expended to manually pollinate each plant. Crops such as cacao beans and vanilla are examples of crops requiring highly labour-intensive practices in the absence of natural pollinators. Throughout history, farmers seeking to grow crops usually confine themselves to growing only the fastest and most productive plants. Such practices can result in growing crops without the genetic diversity found in wildlife. Without such diversity in the genes, crops may become more susceptible to disease and crop failure.

The Great Irish Famine is a well-known example of the dangers of monocultural and mono-varietal crop cultivation. Many scientists, farmers, and businesses have debated how to make agriculture farming sustainable. One of the many practices includes growing a diverse number of perennial crops in a single field, each of which would grow in separate season so as not to compete with each other for natural resources. This system would replicate the biodiversity already found in a natural environment, resulting in increased resistance to diseases and decreased effects of erosion and loss of nutrients in soil.

Nitrogen fixation from legumes, for example, used in conjunction with plants that rely on nitrate from soil for growth, will allow the land to be reused annually. Legumes will grow for a season and replenish the soil with ammonium and nitrate, and the next season other plants can be seeded and grown in the field in preparation for harvest. This method is considered to require a minimal amount of outside resources. In practice, there is no single approach to sustainable agriculture, as the precise goals and methods must be adapted to each individual case.

There may be some techniques of farming that are inherently in conflict with the concept of sustainability, but there is widespread misunderstanding on impacts of some practices. For example, the slash-and-burn techniques that are the characteristic feature of shifting cultivators are often cited as inherently destructive, yet slash-and-burn cultivation has been practiced in the

Amazon for at least 6000 years; serious deforestation did not begin until the 1970s, largely as the result of Brazilian government programs and policies.

There are also many ways to practice sustainable animal husbandry. Some of the key tools to grazing management include fencing off the grazing area into smaller areas called paddocks, lowering stock density, and moving the stock between paddocks frequently.

Urban Planning

There has been considerable debate about which form of human residential habitat may be a better social form for sustainable agriculture. Generally, it is thought that village communities can improve sustainability in that such communities tend to provide a cooperative environment that supports farming.

Many environmentalists pushing for increased population density to preserve agricultural land point out that urban sprawl is less sustainable and more damaging to the environment than living in the cities where cars are not needed because food and other necessities are within walking distance. However, others have theorised that sustainable ecocities, or ecovillages which combine habitation and farming with close proximity between producers and consumers, may provide greater sustainability.

The use of available city space for cooperative food production is another way to achieve greater sustainability. One of the latest ideas in achieving sustainable agricultural involves shifting the production of food plants from major factory farming operations to large, urban, technical facilities called vertical farms.

The advantages of vertical farming include year-round production, isolation from pests and diseases, controllable resource recycling, and on-site production that eliminates the need for transportation costs. While a vertical farm has yet to become a reality, the idea is gaining momentum among those who believe that current sustainable farming methods will be insufficient to provide for a growing global population.

6

FRESHWATER RESOURCES

Earth is rich in resources. Ups and downs in human life depend on the knowledgeable ways in which the resources are used. Of all the resources, water is the most significant. For our ancestors, water was sacred. They believed and treated water better than their mothers. 'Do not abuse water, even if you do so your mother' is an old saying, manifesting this belief. But, in today's society, awareness of the importance of water and the need to manage it have been on the decline.

In the circumstances of today, our needs are based on the prevalent economies. In satisfying these compelling needs, there emerge amongst us certain ideas which are alien to human nature. Hence, every one acts in accordance with the needs of the self and in the development interest of the self. The deeper meaning of life, values and uses are discarded and we act exclusively in the interest of economic gains that can be made. It is for this reason we are generally referred to as the 'homo economicus'.

As for water, there is a widespread 'economic thinking' amongst us. It is because the water is inexpensive and/or free. As such, there is a tendency amongst us to use it in the ways we think best or fit. The world and the users value water inasmuch as it is an economic commodity that must be taken the best advantage of. We act in greed and in the interest of improving the economic well-being of the self/ individuals. There is a selfish attitude in the use of available waters and with a lethargy that does not speak well of us. As such, the water is contaminated and is wasted by pollution. Consequently, the quantity and the quality of water have been declining fast.

FRESHWATER

Freshwater is defined as having a low salt concentration — usually less than 1per cent. Plants and animals in freshwater regions are adjusted to the low salt content and would not be able to survive in areas of high salt concentration. There are different types of freshwater regions:

1. Ponds and lakes,
2. Streams and rivers,
3. Wetlands.

Access to unpolluted fresh water is a critical issue for the survival of many species, including humans, who must drink fresh water in order to survive. Only 3 per cent of the water on Earth is fresh water in nature, and about two-thirds of this is frozen in glaciers and polar ice caps. Most of the rest is underground and only 0.3 per cent is surface water.

Freshwater lakes, most notably Lake Baikal in Russia and the Great Lakes in North America, contain seven-eighths of this fresh surface water. Swamps have most of the balance with only a small amount in rivers, most notably the Amazon River. The atmosphere contains 0.04 per cent water. In areas with no fresh water on the ground surface, fresh water derived from precipitation may, because of its lower density, overlie saline ground water in lenses or layers.

VALUABLE RESOURCE

Water is not just an economic commodity. It is our life. In the use of it, we must not act in the self interest, and with a narrow objective of satisfying the need of the self alone. We must safeguard the water resources in a way it will be useful for the posterity and others in a sustained manner. Its quality has to be protected. If we act keeping in view the twin focus of water quality protection and use by the generations of the future, then it would be in line with the sustainable use of the resource. It is in fact in this sense that water is considered the lifeblood of the earth.

Hence, in the use of the liquid named water, we should keep in mind the following:

1. Water is valuable resource.
2. Although it is obtained inexpensively or as a free resource, it has high economic value.
3. In its use, there is no place for selfishness, greed and the narrowminded approach of 'self development'.
4. As it is increasingly becoming scarce, there is need for greater attention in its use.
5. Water quality must be protected. The individuals and the people have an equal responsibility.
6. It is good to understand others' thinking and act in an integrated manner in the management of water resources.

It is therefore important that we act in the knowledge that water is a valuable and significant resource.

QUANTITY AND QUALITY OF WATER

Sources of Water: Earth scientists speak of two ideas for the appearance of water on the surface of the earth.

1. There was volcanic activity on the earth several million years ago. Hydrogen and oxygen escaped from the volcanic activity. The two then joined to give rise to water.
2. Comets had been entering the earth's atmosphere over several million years. The ice carried by the comets evaporated and became water vapour. This water vapour had then condensed to fall as rains to the surface.

The two activities go on in the other planets as well. Yet, there are no water bodies in these planets because the atmosphere that envelopes the earth is not found in these planets. The earth's atmosphere captures the water vapour. The temperature regime of the atmosphere helps to keep most of the water vapour in the liquid form. Hence, there is opportunity for the development of water on the surface of the earth. It is primarily for this reason, our planet earth is known as the 'water planet'.

Quantity of Water

Water is the earth's foremost problem. Of the earth's water, 99 per cent is not available for direct use by humans. Salinity and the form in which it is available and location, glaciers, ice caps are the primary reasons for such a state of affairs. For 1.0 per cent of the world's water, the entire humanity is in competition. More than 97 per cent of the water on the earth is salt water. This is not directly used by the humankind. The remaining 3 per cent is freshwater.

The quantity of water directly available for human consumption is 0.014 per cent to the total. The rest is found in some seas, lakes, rivers, glaciers and icecaps as well as underground. Surface water is about 230,250 ckm, of which freshwater lakes hold 125,000 ckm, saline lakes and backwaters 104,000 ckm and rivers and fountains 1,250 ckm. Groundwater is about 8.41 million ckm, of which soil humidity is 67,000 ckm, sub-groundwater and deep water 417,000 ckm.

While the groundwater accounts for less than one per cent of the total water, the surface water accounts for a negligible proportion only (0.0171 per cent). Ice and glaciers account for 2.15 per cent of the total waters (or 29.18 million ckm) while atmosphere for 0.001 per cent (13,000 ckm).

Freshwater Sources: Most freshwater is locked up in ice. It is stored also as the groundwater in great quantities. Only the remaining fraction of freshwater is directly available to us as the useful part of the freshwater from the lakes and the rivers. These water bodies are discussed below:

1. *Icecaps and Glaciers*: Of the world's freshwater, 77 per cent is in solid form in the icecaps. The poles are covered by the icecaps. Antarctica and Greenland are entirely covered by ice. A portion of these flow as the glaciers. Thus, the freshwater is found locked up for a thickness of several thousand metres in the ice covering the polar regions. In these regions, the precipitation is also in the form of snow. As the ice does not melt even during the summer in these parts, the ice becomes solid and becomes the rocks of ice and icebergs. About 29 million cubic kilometre of water is in the form of ice.
2. *River Flows*: The surface waters flowing in the rivers are the most useful for us. The waters of the rivers are received primarily from precipitation.

Sometimes, the rains gather as snow or ice and turn into surface waters in summer or spring seasons. About 60 per cent of the total geographical area of the world is made of river basins. There are several large rivers such as the Amazon and Congo draining the equatorial region. In the South and Southeast Asian regions, there are equally large rivers as those of the Indus, Ganges, Hwango and Irrawady. There are no large rivers in the temperate region as are found in the tropical areas. Yet, because of the special climatic features of this region, the rivers here are perennial. Examples are the Thames, Rhine and Volga. Due primarily to local conditions, streams that run once in several years are formed in the tropical deserts.

A majority of the people live depending on the river waters. But, the quantum of water in the rivers changes with the seasons. As a result, for some months in a year, there may be water scarcities. Groundwater comes in useful during these months.

3. *Underground Water*: In dense forest areas, much groundwater is gathered. The forests of the tropics, temperate areas and also Taiga regions possess much groundwater. In semi-arid regions also the groundwater is gathered due to rains. Such groundwater aquifers are found in Australia, in the provinces of Queensland, New South Wales and Victoria. Water emerges from the rocks beneath the surface as artesian springs. Groundwater is also found in abundance in the coastal areas. During the times of scarcities, the groundwater of the coastal area is greatly used by the people living along the coast. At times, the groundwater is so very highly exploited that the sea water intrudes down below.

As a result, the wells along the coast yield salt water. In most areas of the city of Chennai, there is such a situation.

4. *Water Recharge*: All the freshwaters mentioned above can be recharged. Yet, the water withdrawn is much more than what is recharged and hence over the years these sources become dried up. When the quantum of water declines in the water bodies, there generally arises a scarcity. With scarcity, the quality also goes down. Particularly, the freshwater loses its quality through contamination and pollution.

Quality of Water

When the humans began to settle down permanently in some areas, the waters in those areas declined in quality through their economic activity. Increasing population, agriculture, industrial activities and urban needs make a large demand on the freshwater sources. With the changing climate, there is a general decline in rainfall as well. Besides these, bio-degradation, salination through irrigation and such other events cause declines in quality. Nowadays, the wastes or effluents disposed from the industries such as the solid wastes, heavy metals, radioactive materials, nitrates and minute carbon pollutants create problems in water quality.

Acidification of the waters of the lakes and streams and declines in oxygen content in the coastal waters are expected to create major problems in water quality in the near future. We have now seen how the water quality declines due to various reasons. Our life is intertwined with the protection of the water resources. A quality assessment of the freshwaters made by the World Watch Institute has indicated to the declining quality worldwide.

This assessment has also indicated to improving quality of the freshwater in some of the industrialised countries. It is pointed out that the quality of freshwater and the protection to it are still not adequate worldwide. The controls remain weaknesses. Just as quality, quantity has also become a worldwide problem. Increasing population further compounds this problem. Water scarcity is likely to become a Crisis of the Future.

Water Based Diseases

Health of humans is affected because of water pollution, quality deterioration, scarcities and water caused diseases. Diseases that affect human may be conveniently classified as five types, following the World Health Organisation, as shown below.

1. Water-breeding insects such as mosquitoes carry malaria, filariasis and yellow fever. Blackflies transmit river blindness via a parasitic worm. Water-breeding diseases: malaria, filariasis, yellow fever, blackflies and river blindness. Malaria carried by mosquitoes, affects perhaps 160 million people at any one time. There are an estimated 800 million infections each year.
2. Water-washed Diseases such as scabies, trachoma and leprosy and conjunctivitis attack the skin and eyes. All are spread by insufficient water for personal hygiene. Water-washed diseases are scabies, leprosy, conjunctivitis and trachoma.
3. Water-based Diseases carried by invertebrates. Schistosomiasis is transmitted by snails; guinea worm is a parasite of a crustacean water flea. Schistosomiasis has been linked to spread of irrigation canals and reservoirs, a veritable habitat for snails. 600-1000 working days are lost per case.

4. Waterborne Diseases are spread by drinking or washing in contaminated water. Diseases include typhoid, cholera, dysentery and diarrhoea. Diarrhoea kills 6 million children every year. Repeated attacks exacerbate malnutrition, reducing food intake and impairing absorption.
5. Defective Sanitation aids the spread intestinal worms. Eggs are excreted in human faeces. The infection cycle begins on swallowing the eggs. Hookworm larvae usually enter humans by burrowing through the soles of the feet. Heavy infestations can cause death in children.

Trachoma is a contagious inflammation of the inner lining of the eyelids. This often causes blindness. Some 500 million people in the world suffer due to the severity of these diseases. In the countries of the North (developed), 9 out of 10 people have ample clean, piped water and sanitation. In the South (developing and underdeveloped), only 2 out of 5 people have easy access to safe water and 1 in 4 to proper sanitation.

The provision of clean and adequate supplies of water is one of the major problems of humanity in the developing world today. Rivers, lakes and ponds serve both as a source of clean water and as a sink for all wastes generated by humans. Dirty water is believed to be the principal transmission agent for at least 80 per cent of the diseases which afflict the developing countries.

The rural poor, who lack access to standpipes, have little choice but to collect their water from rivers, streams, ponds, mud holes and wells. Millions of women and children often spend 6 hours a day walking long distances to bring home a few litres of dirty water. Water scarcities thus call for a concern for water supply and overcoming scarcities of any type.

SCARCITY IN DRINKING WATER

Scarcity in drinking water indicates not only to the per capita water availability but also to the difficulties in obtaining the water. In the developed and industrialised countries, water is supplied largely through taps. But in most developing countries people have to walk to a source for a minimum of 2 km. The water needs of three different states/countries of varying economic development are shown. It seems that the drinking water scarcity is seen as a consequence of increasing population numbers. This is however true only to a certain extent. There are indeed several reasons for the drinking water scarcity. Most important of these are urbanisation and industrialisation.

Urbanisation

Villages turn into towns, towns into cities and cities into metropolises, in course of time. Towns develop industries and generate employment. As a result of this, people from villages go into town, in search of jobs. Thus rural to urban migration occurs. The basic needs of the urban population increase not only as a result of natural increases in population but also due to migration of people from one place to another, especially to the towns and cities. In particular, scarcities are created in the fundamental infrastructures. It is because there has not been any concerted effort at improving supply of water.

Water scarcity is thus compounded by other problems, importantly by not improving the supply position. Drinking water scarcity becomes acute because of this. In the world today, only in a small number of cities, water sources and distribution are found commensurate with the growth of the city. When plans are made and implemented, the population overshoots the development. For example, in the metropolitan cities like New Delhi, Mumbai, Kolkata and Chennai, scarcities of water are high due to the population growth. The inability to store water at the required levels is yet another reason for scarcity in the drinking water.

Drinking water for Chennai has been stored in the suburban lakes such as the Chembarapakkam, Puzhal (Redhills) and other sources. These storages are not adequate enough for storing water needed to the city of Chennai. It is because there has been a high siltation in these sources. This has considerably reduced the quantum that could be stored at any given time. As a consequence, the surface runoff as a result of rainfall flows very quickly into the sea. It is because of this reason that water has been transported over large distances, from such places as Neyveli, using the trucks and railways.

The expenditure incurred in recent years in the transport of water from far off places has mounted to Rs.1,000 million a year. People from rural areas continuously migrate to the cities in large number in search of employment, education and other reasons. The streams that migrate to the cities find their housing in the slums because there is a general shortage of housing. This further aggravates the shortages in water supply and pressures on existing amenities, causing suffering to the people. With congestion in cities, even the most carefully planned water supply programmes are unable to provide sufficient water supply.

Almost every city, in the country, has water scarcities in some months of the year that the drinking water scarcities remain a permanent crisis. Only in a small number of villages, there are quality water sources. In many Asian countries, there is no domestic or drinking water supply by individual households. In the rainy seasons, the water that flows in the rivers becomes the source for both.

When the river goes dry, the wells, ponds and lakes also go dry. Hence, scarcity exists until the next rainy season. Most peninsular rivers of India are of non-perennial nature. It is for this reason that the water is impounded in dams built across the major rivers. In dry periods, water is let into the rivers from the dam. In times of monsoon failures, the dams also go dry. Industrialisation and other related developments lead to contamination of the freshwater sources.

Industrialisation

Whether town or village, the main cause of water scarcity is industry. In the towns, the water supply has to satisfy both drinking and industrial uses of water. The increasing industrial units have always, for reasons of solvent, coolant and as a source resource for production, increased the demands. With the industrial demands increasing, there has been a short supply in the residential areas of the cities and towns.

A particular reason for urban water deficits is that there is a limited opportunity for storing water in them. This is so because the cities are becoming entirely built-up areas, with no provision made for adequate storage construction. Where such constructions exist, the sources of supply have either become dry or brackish, necessitating the fetching of water from far off places. Such external sources are not protected and hence the water is not good quality.

At the global level, there is an increasing decline in water quality. It is difficult at the same time to make measurements and gather quality related data. This causes problem for determining the quantum of freshwater required for a given population. The national governments of the world countries have made efforts towards regulatory procedures and protection.

LAND AND WATER POLLUTION

Water resources are polluted by the wastes generated by the humans. Water in ponds, lakes, tanks, rivers, seas and oceans and in all other forms gets polluted by the human activities. On the one hand, there is a scarcity of water while, on the other, there is pollution of it. Polluted water affects land and degrades it. Now let us look at how and why water pollution and land degradation occur.

Water Pollution: Water pollution in the world is becoming a serious crisis. There are multiple causes for water pollution. They are:

1. Water draining from agricultural lands, carrying chemical pollutants;
2. Wastewater disposal in urban areas;
3. Industrial wastewater disposal;
4. Wastes disposed of and temperatures of wastes from industrial units such as atomic and other energy production units; and
5. Pollutants of oceans.

Wastes from human activities are the prime causes of water pollution. The indiscriminate dumping of wastes and the natural drainage of the water cause pollution to a greater extent.

Agricultural Wastewaters

Modern agriculture uses chemical fertilisers, insecticides and pesticides. Fertilisers broadcast in the cropped fields, insecticides and weedicides dusted on to the plants dissolve in the water—from rains and irrigation water—and are drained into canals and rivers. They finally reach the sea. It has been estimated that 83 per cent of the sea and ocean pollutants have human activity origins.

Urban Wastewaters

With urbanisation, scarcities in basic amenities have become the order of the day. There appears to be no plan for proper disposing of wastewaters from urban activities. Due to this shortcoming, wastewaters find their way to water bodies. In Chennai city, for example, the rivers such as the Cooum and Adyar carry domestic wastewater. Industrial also reach these and create a foul smell in the air. The rivers which were once the quality freshwater bodies have now become urban sewers.

On the contrary, rivers such as the Thames of London and the Seine of Paris which were once much polluted and have now been cleaned up. The Ganges was also a heavily polluted river not long ago and has now been cleaned under the Clean Ganga Project. Yet, there are several water bodies in the cities and towns of the country which carry wastewaters. The hazardous chemicals which are carried in solution and suspension finally end up in the seas, causing water pollution there too.

Industrial Wastewaters

Hazardous wastewaters from industries are also drained into the rivers and reach the oceans. As a consequence, the recycled water from the hydrological cycle is also a pollutant and contains contaminant traces. Food wastes, metal wastes and chemical wastes are not easily degradable. Hence, their traces are amplified over the years. Equally disturbing is the fact that radioactive wastes and high temperatures from it are transferred to the water bodies.

Atomic Wastes and Oil Spills

The countries which produce atomic energy claim that the production is for peaceful purposes but they dump the radioactive ashes along the coast, marsh lands and estuaries. This causes the temperature of the sea waters to rise. Consequently, its quality declines. An estimate has it that nearly 200,000 tonnes of oil is lost as leaks from the industries into the water bodies. Oil spills in the oceans greatly affect the sea organisms, fish and sea animals.

Human beings are affected when they consume sea food, including fish. Another industrial by-product that pollutes water bodies greatly is the tannery effluent. It is often indiscriminately disposed of. In the tanneries, every 100 kg of skins and hides processed requires 3,000 litres of water. This water is used for soaking and is used in almost all processes of tanning.

Almost the entire quantity of water used in tanning is disposed of as wastewaters. In places such as Vaniyambadi, Ambur and Ranipettai in North Arcot-Ambedkar, there are hundreds of tanning industries. Wastewaters are not treated before disposal. They are disposed of in the open and into the river Palar. The Palar irrigates lands in its command and thus pollutes the land as well. Even the few wastewater treatment plants in operation in the district are not functioning properly for reasons of power shortage and excessive flow of water.

Land Degradation

Water pollution often leads to land degradation. Water infiltrating from the wastewater sources reaches the underground aquifers and contaminates the underground water. Land is degraded wherever surface waters and drainage and sewage channels carry wastewaters and wherever the sewers are either clogged or broken. Land resources of the urban areas are degraded due to indiscriminate solid wastes and wastewaters dumping in the open. In cities such as Chennai, Salem and Tiruchirappalli in Tamil Nadu, visual blight is appreciable.

In the rural areas, however, such blight is hidden and thus are not immediately apparent, except where it is widespread. There are of course other types of land degradation. In the Kambam valley of Madurai and/or Theni districts, there is a stretch of land along the Western Ghats fully covered with sand dunes and sand encroached areas, known popularly as the 'theri sands'. This stretch is 120 sq. km in extent and therefore land degradation is over a vast area.

Although the irrigation water from the wells, rivers and canals is contaminated by solid wastes and wastewaters, chemicals such as fertilisers, pesticides and insecticides, there is only limited awareness of the degradation it causes in the rural areas. Worse, the farmers and the people are not even aware of it. Researches have shown the levels of pollution in such areas through careful measurements.

As a consequence of the events described above, negative impacts have now surfaced. These impacts have pushed us to the options of land and water conservation and management, and importantly earth's resources and its environment. There is a need therefore to work towards a sustainable future. This future has to be achieved through appropriate environmental and resources management. In regard to land and water, there has to be an integrated management, with an ethical framework.

ETHICS AND MANAGEMENT OF WATER

Water resources are being used in several ways towards world development. Water scarcity and decline in water quality have been the end result of economic activities which use water. We have understood the reasons behind the scarcity and decline in quality as those of a high demand for water, improper water utilisation and the progressively increasing population. It is the turn now to know and understand the ethics behind the use of water and water resources management.

Water Ethics

Much as we use water for our daily activities, there is a certain moral responsibility on our part to use it and this is the water ethics. There are in fact two ways in which water ethics could be assumed and taken care of.

1. It is a moral responsibility for all of us to use the water resources without much waste and contamination and realising them in the light of the fact water is available free and in plenty. There is a need to use it in quantities absolutely necessary than waste it. It is important also to be concerned with the needs of others who have the compulsion to use it just as we do. Distribution with justice is in effect good water ethics.
2. We should not be the perpetrators of water contamination nor should we be the silent spectators to the acts of violation of water ethics by others.

It follows from above that, in short, all our water related actions should always be such that everybody benefits from its use, in an ethical way. This way we all should be able to take an appropriate role in the water management.

Water Demand

Water is a renewable resource. But in the next 15 years, water scarcity will be very high in 30 countries of the world. It has also been estimated that, by the end of the century, the water need would either double or treble the quantum of need in 1980. It is possible to satisfy the needs of the world population, if only we could follow a careful water management strategy. There are opportunities also for satisfying the need in excess of demand.

Irrigation of Water

Irrigation Department is a major user of water. Of the global water use, 70 per cent is accounted for by irrigation. With irrigation, 12 per cent of the world's crop lands obtain benefits. There are chances for the crop lands to double by the year 2000 A.D. This means that irrigation might demand greater water use in the near future. Water use in a country depends very much on its efficiency in water use. Even in most technologically advanced country, the efficiency of irrigation is only 30 per cent.

As the people of California, in the United States of America, and Israel use water ethically and with enormous care, the efficiency in use is high. They are also good in water management activities. For example, drip irrigation has been used in Israel as well as California to best effect. The farmers and horticulturists make the water drip drop-by-drop at the roots of the crop plants. This system of irrigation has now come into use in India too.

With this method of irrigation, a high output of crops is being harvested with small amount of water. In this way, salination of the lands is also averted to a great extent. Israel is an exemplary example of water use. In this country, 95 per cent of the water resources has been brought under use.

Desert has become a flourishing agricultural land. A fifth of the industrial and domestic waters are treated and then used in irrigation. In 1962, 20 cubic metres of water was required to produce US $ 100 worth of goods. In 1975, it took only 8 cubic metres of water for the same level of production. And today, Israel uses much less water than 8 cubic metres to produce goods worth more than US $ 100.

Political Dimension

The one area we have to tread carefully in water ethics is that of the political dimension of water use. Today's political climate is such that water use is opposed to ethical uses. More than 75 per cent of the World's 200 major rivers are shared by two countries each. India and Bangladesh are entangled in a dispute involving the Gangetic waters while Brazil and Argentina are at dispute involving the La Plata.

In the world, there have been constant and continuing disputes between countries in respect of 6 major international rivers. It will be beneficial to approach these disputes through mediation and cooperation. A river basin is a natural region. The basin does not place importance on the national boundaries. But 'the basin is a resource' is an important value. In the countries the Nile and the Danube flow, water is considered a valuable resource. As for the Nile, Egypt and Sudan reap enormous benefits. Egypt carries out its agricultural and industrial activities in an integrated manner.

Water Management

In today's circumstances, we need a global management of water. World water demand is of course less than world water supply. Yet, water demand from sectors such as agriculture and other industries has been increasing very rapidly. The problems that emerge from the situation therefore indicate to the nature of water demand. There will be water scarcity at least in 30 countries the world over by the year 2000. Hence, to improve the availability of water and also the quality of water, proper management is a necessity. There are at least two significant means in water management that need to be looked into. They are:

1. Investing on the activities that help store water from the hydrological cycle and related activities such as dams; and
2. Demand management is making efforts to get water supplied where it is needed.

Building dams across the rivers try to satisfy multiple demands at the same time. Flood control, improving hydroelectricity generation and other uses of water including of irrigation are all included in this activity. The dams are also useful as multipurpose resources, freshwater fish, prawn cultivation and recreation resources. For example, the Aswan Dam in Egypt has been a good effort at national development.

It provides for 50 per cent of the national demand for electricity and also gives protection from the floods. There are also negative impacts. There have been, for example, heavier clayey and sandy sediments, amounting as much as 100 million tonnes, deposited in the command area. The dam which once made the lands of the command very fertile causes sediments in the Nasar Lake.

As a result, to provide nutrients to make the lands fertile enough for production, the import of fertilisers has been on the increase. Besides, the brick kiln operators of Cairo do not get adequate raw materials for their brick production. It has also been found that the nutrients that were supplied to the sea from the Nile have now become scarce or even rare. The delta has gradually been receding over the years.

Land salination and water-logging have increased, causing a degardation of land. The Food and Agricultural Organisation (FAO) of the United Nations has estimated that 35 per cent of the land in Egypt has been affected by salination while 90 per cent of the land in that country has been water-logged. This brings us to the fact that in water management, basin management is the foremost in importance.

Basin Management

In half the countries of the world, the demand for water would double by the year 2000 due to population growth. The pressures emanating from this fact would hasten the problems of the basins. This impinges on other resource problems in our lives. It is therefore important that the river water is a significant part of the water management efforts, too. This is also important for the reason that a problem affecting one part of the basin would automatically affect other parts of the system; an event occurring at one end of the river would impact upon the other end of the river.

River basin, like other resources, should be shared. Hence, water management through participatory approach would provide benefits for all people in the basin. Just as there are conflicts in 148 of the 200 major river basins, there are opportunities in all of these for integrated efforts at management. As of now, however, several countries have transformed their problems into crises. At the global level, there are only a few countries which cooperatively and understandingly approach the problem for solutions. From the experiences and successes of these countries, participatory approach emerges as the most effective for resolving the problems.

7

GLOBAL FOREST WEALTH

Forests have a variety of uses to humans, including wood from trees, nutrition from animals, for grazing, recreation, medicinal and so on. At the present time, conservationists are still arguing about a 'technical' definition of a forest. According to The Food and Agriculture Organisation of the UN (FAO), a forest does not stop being a forest just because the trees are gone. While that may be so, it is important to understand how the disappearing green cover and the resultant threat to habitats and to human life fits into the bigger picture of life on the planet.

The forest is a complex ecosystem—a biological system with distinct, myriad interrelationships of the living part of the environment (plants, animals and microorganisms) to each other and to the non-living, inorganic or abiotic parts (soil, climate, water, organic debris, rocks). Picture it as an intricate web - fragile but at the same time holding the ecosystem together.

TYPES OF FOREST

Forests come in all sizes and types—from the northern taiga to the scrub forests of arid regions to the rainforests of the humid tropics. They are found on moving glaciers, in fresh and salt water, on arctic mountain slopes. They do not occur in isolation from the rest of the landscape. The type of forest in a given area depends on many elements, including climate, soil, water source, rainfall patterns, seed sources and human influence.

The complex ecological relationships involving forests could allow humans to benefit from them in a variety of ways. However, a deeper understanding of these relationships is crucial for development of effective forest management and policy options.

IMPORTANCE OF FORESTS

Use of Forest Products

All these and many more activities directly or indirectly involve forests. Some are easy to figure out—fruits, paper and wood from trees, and so on; others are more difficult —by-products that go into the manufacture of other everyday items like medicines, detergents, etc.

Biodiversity and Livelihood for Humans

But looking at it beyond the narrow, human, not to mention urban, perspective, forests provide habitats to diverse animal species, and they also form the source of livelihood for many different human settlements as well as for governments. They offer watershed protection, timber and non-timber products, and various recreational options. They prevent soil erosion, help in maintaining the water cycle, check global warming by using carbon dioxide in photosynthesis.

The entire ecosystem begins to come apart, with dire consequences for all of us. Thus, it is quite scary that over the past fifty years, about half the world's original forest cover has been lost, the most significant cause for that being humans beings' unsystematic use of its resources. It brings to mind Francis Bacon's words about there being enough for everyone's need, but not for anyone's greed.

Earth's Air Purifiers

Just as the lungs absorb carbon dioxide from the blood and infuse it with oxygen, green plants absorb carbon dioxide during photosynthesis and release oxygen into the atmosphere in return. Which is why forests are often referred to as the Earth's lungs. This epithet is most widely used with regard to the Amazon rainforests, the world's largest surviving tropical forest.

Carbon Guzzlers

Photosynthesis—the production of energy in the presence of light by chlorophyll - containing plant parts for the subsistence of the organism. Carbon dioxide, the major greenhouse gas, is a major requirement of photosynthesis. This suits us just fine as it means forests form an effective sink for the carbon dioxide produced as a result of animal respiration, burning of fossil fuels, volcanoes and other natural and human-induced phenomenon. And if that is not all, a by-product of photosynthesis is oxygen. Thus, the Amazon forests are the Earth's air purifiers, given the large amounts of carbon dioxide they absorb from the atmosphere.

However, these South American rainforests—all forests, really—are being depleted at alarming rates. The lungs of the planet are increasingly being likened to those of a compulsive smoker. Scientists, including the FAO, estimate that more than 50 per cent of the original rainforest cover has been lost since the beginning of the twentieth century, and what remains is disappearing at an alarming rate. Many of these forests have existed for millions of years, and over a third of the world's plant and animal species live here. In some areas more than 200 tree species have been known to exist on a single hectare of land!

Logging and Accidental Fires

Deforestation had been the major culprit in the disappearing forest cover. However, in a study published in Nature in late 2004, researchers say that the forest is disappearing faster than we think, as logging and accidental fires—which were assumed to have minimal impact—are not taken into account in standard calculations. In the Amazon, they said, logging practices and wild fires destroy as much or perhaps more annually than deforestation, and this affects more communities than earlier estimated.

Forests as Habitats

Although the ocean was the original home of all life of Earth, forests, as they themselves evolved, quickly became home to a vast majority of land based creatures including early man.

Natural Home to Biodiversity

Today, the vast amount of diverse life that can be found in forests is evidenced by the startling statistics that accompany any analysis of a rainforest. Although only covering 6 per cent of the planet's surface, these lush green, often tropical masses contain around 50 per cent of plant and animal species on this planet. In one square kilometre of rainforest, you can often find more types of life than can be found in an equivalent 1000 square kilometres in colder, more northern climes.

All forests are great collectors and storerooms of water. Their root structure holds together the soil, and their leaf litter gets broken down and combined with minerals to form the equivalent of gigantic sponges—slowly releasing water into surrounding areas at a dependable rate. A forest is home to many types plants, which are the food source for many types of animals, which are, in their turn, also sources of food for other animals. And, as these animals and plants die, they in turn become food sources for the plants that again become food sources for the animals.

This circle of life, the linkages between all animals and all plants, is often referred to as "The web of life"—a reference to the common dependencies between all life in an ecosystem.

Shelter is the last of the triumvirate. Trees take on the worst (and the best) of the elements: wind, sun, rain, temperature. By taking the brunt of these elements, under their cover they can lessen the impacts of too strong a sun, of destructively heavy rainfall, of lessening the vagaries of temperature change, and taking the punch out of strong winds. In providing so many benefits, a plethora of both plants and animals alike have adapted to and taken shelter underneath the protective canopies of trees.

Consider so much life depends on and lives in forests, the fact that forest cover is now a fraction of what it used to be even a few hundred years ago, it stands to reason that there is less space for this diversity of life to live in, and increasingly greater contact with encroaching humans.

Animals that normally live within the boundaries of a shrinking forest are forced to come out and look for food. This often brings them into conflict with human settlements that are often situated near forests (because forests provide food and water).

Humans and Forests

Concrete jungles, factories spewing smoke, crowded roads crammed with traffic—certainly, humankind has come a long way from its humble beginnings. Plants themselves have been around for over 450 million years, starting with simple forms of single—layers of cells. But today they are among the most complex of life systems. And also the oldest and most successful. Imagine - the first plants stood upright about 420 million years before the first animals could!

Human Evolution

It is widely believed that early human species evolved in and around the African rainforests between four and six million years ago. Even today, the closest relatives, the great apes, live there—one of the most productive, if difficult, ecosystems on the planet. Recent fossil discoveries, however, pin the figure closer to four billion years, when the first bipedal ancestors of Homo sapiens made the wooded grasslands of East Africa their home. The forest offered them food in the form of plants and animals, and water. After the discovery of fire, wood was used for burning, and later on for making tools. As life evolved, they started clearing land for agriculture and to set up more and more advanced settlements.

Human beings and forests have always had a complex relationship. We have depended on forests as long as we have inhabited the planet—getting clean air to breathe, food and water from it, fuel, shade and shelter, and now we need it for economic gain as well.

Early humans were known to worship trees, and even today, in some parts of the world, forests are regarded as places of awe, with spirits attributed to be living there. The worship of forests, plants and animals, and appeasing of animal and tree spirits are still quite common in some cultures, and the forest is treated with the kind of respect reserved for divine objects.

Forest-People Equation

Today, the ecosystem that gave us life is under severe threat. Early humans were hunter—gatherers, and later farmers, and because the population was small, the impact on the environment was minimal. But in the past couple of thousand years, the growing demands of an ever-increasing human population has halved the Earth's original forest cover. Great Britain was almost completely covered in forest at one time - today, it has a forest cover of around 10%!

Forests are the Earth's largest and most productive ecosystem, and trees their most visible and important constituent.

ECONOMIC BENEFIT OF FORESTS

It is estimated that the forest products contribute about 1% of world gross domestic product (GDP). The annual turnover of timber and other wood products from forests is valued at more than US$200 billion. Apart from that, non-timber products like rubber, cotton, medicinal products, food and so on represent significant economic value. Even more important is fuel wood and fodder, especially in developing nations, where people depend on wood almost entirely for their household energy. Given the immense economic benefit of forests, the demand for commercial timber and other products is ever increasing, and expected to rise by 50 per cent by 2010. Already, there are signs growing shortage of tropical hardwoods.

A Lessening Resource

While over-harvesting of timber is usually identified as the main cause of deforestation, it is just one of the many reasons. Increasing human population and the need to feed and sustain them; agriculture, including the demand for cultivation of cash crops; mining; water storage; and urbanisation and related activities shoulder much of the blame. As for fuel wood, it is estimated that by 2010, most of the 3 billion people who depend on it for their daily living will find it hard to obtain. Already, rural families spend precious hours in collecting firewood instead of other productive work, something that causes losses to the tune of US$ 50 billion to the world economy.

Forest Certification Process

However, it has been proved that it is possible to harvest timber and other forest products in a sustainable fashion, without disturbing the forest ecology at large. It is in the interest of the world's timber trade (not to mention the environment!) that forests are managed responsibly, and trade in forest products from threatened species is outlawed. This is where forest certification comes in.

Forest certification is a two-stage process. First, forests are independently certified to a recognised standard, such as the Forest Stewardship Council (FSC). Next is the certification of operations in the timber supply chain, referred to as the 'chain of custody' certification. Basically, it is a simple equation. If the annual cut in a forest exceeds the annual growth, it spells trouble. However, good management policies, good product design and minimum waste can be used to safeguard the resource for the future.

Types of Wood

The terms hardwood and softwood are often used in relation to timber. These refer not to how soft or hard the wood is, but to the structure of the trees. In fact, certain hardwoods, for instance, poplar, are 'softer' than softwoods like pine!

Hardwoods

The scientific term for a hardwood tree is angiospermae. This means the plant encases its seeds in seed cases. They are of two types, monocotyledoneae (whose seed cases have a single lobe) and dicotyledoneae (two lobes). Hardwoods are generally broad-leaved trees, and are deciduous.

Some hardwoods and their uses:

1. Mahogany Quality furniture such as cabinets; boat construction; wood facings and veneers.
2. Walnut Gunstocks, solid and veneered furniture, novelties, cabinetry and wall panelling.
3. Oak Furniture, trimming, boat framing, desks and flooring.
4. Maple Flooring, fine furniture and wood - ware such as bowling alleys.
5. Cherry Cabinet making, boat trim, novelties.
6. Rosewood Musical instruments, piano cases, tool handles, art projects, veneers and furniture.
7. Teak Fine furniture, panelling, shipbuilding, doors, window frames, flooring and general construction.

Softwoods

Softwoods are generally coniferous, evergreen trees, found in colder, northern climes. They are known as gymnospermae, which means their seeds are 'naked', and not covered by a seed case or pod. Coniferous softwoods form the major part of the world's commercial timber.

Some important softwoods and their uses:

1. Pine House construction, panelling, trim, furniture, moulding and boxes.
2. Hemlock Construction lumber, planks, doors, boards, panelling, sub-flooring and crates.
3. Fir Furniture, doors, frames, windows, plywood, veneer, general mill-work and interior trim.
4. Redwood Outdoor furniture, fencing, house siding, interior finishing, veneers and panelling.
5. Spruce Masts and spars for ships, aircraft, crates, boxes, general mill-work and ladders.
6. Cedar Chest making, closet lining, shingles, posts, dock planks, novelties and Venetian blinds.

Cellular Structure of Woods

Softwood: Simplified softwood cellular structure showing strength providing thick-walled tracheids on the right. Thin-walled earlywood tracheids are seen at the left. Fin medullar rays run across the tracheids and are used to store food. The large opening is a resin canal.

Hardwood: Marked difference between large earlywood pores and small latewood pores characterise hardwoods. Strength is provided by the large number of fibre cells in latewood. Food is stored in parenchyma tissue laid in rays at right angles to the rings.

Uses of Timber

Timber goes into the making of textiles, tyres, lenses, toothbrushes, diapers, sanitary napkins, newsprint, packaging material, paper products, food thickeners, tissues. Building material like frames, flooring, cabinets, panelling, boats, gunstocks and more all use wood. Sheds, barns, docks are usually made of wood, too.

There are a number of factors to be considered while using wood. In environmental terms, there are both positives and negatives in using wood-based material. The good thing is, timber is a natural product and can be used with minimum processing. It occurs almost worldwide, and is essentially a renewable and recyclable resource. Because of its natural origins, most types of wood are fairly durable. Timber has a high strength to weight ratio that allows for efficient transportation.

On the negative side, the high demand for timber products encourages illegal logging, even though some timber trade around the world comes from certified, well-managed sources. Transporting of timber also adds to greenhouse gas emissions, especially for exotic species that are particularly in demand in places far away from where they grow. Of course, one way to curb this is to encourage use of locally available species.

Also, processing of timber can have undesirable effects on the environment. Dust emissions result from the sawing and sanding of wood-based products. Sometimes chemicals used the treatment of wood may be toxic. Treated timber also raises issues during recycling and disposal. Even so, the processing of other everyday materials such as plastics and aluminium are far more damaging.

Non-Timber Forest Products

Referred to as Non-Timber Forest Products (NTFP) or NWFP, they include food, rubber, resins, gums, medicinal or cosmetic products and more. Broadly, NTFP may be classified as edible and non-edible plant and animal products. The former include edible plants and animals, honey, oils, fish, spices and so on. Non-edible products include grasses, bamboo, ornamental plants, insect products, oils for cosmetic use, medicinal products, etc.

Economic implications

From the economic viewpoint, NTFP are equally important as wood-based products. About 150 types of NTFP are significant in international trade. They are also increasingly being acknowledged for their role in sustainable development and conservation of biological diversity.

Up to 80 per cent of the population in developing countries depend on NTFP for subsistence, both economically and for nutrition. NTFP are especially important to women in developing countries from Latin America, Asia, Africa and Far East. Though conservation agencies are addressing the issue of sustainable production of NTFP, there are a number of challenges to be met, some of which include the disappearing forest cover, inequitable market access of marginalised population and the monopolisation of high-value NTFP by logging and poaching mafia.

OTHER USES OF FORESTS

Recreation Uses of Forests

Forests have always been popular for outdoor activities. Hiking, camping, fishing, sight-seeing, boating are all recreational pursuits for which the forest can be exploited, and if well managed, all of these can be of immense benefit to the immediate community as well as to the forest authorities.

With increasing urbanisation, the demand for outdoor recreational facilities rises, and therefore the need for forested lands near metropolitan areas to be managed accordingly. While natural beauty is enough to attract visitors, forest management authorities have to budget for access roads, camp sites, parking lots, beach facilities, docks and so on with minimal damage to the forest itself.

In many parts of the world, state forests have regulations regarding hunting and fishing. Even private forest owners acknowledge that opening the forest for recreational activities means not just economic benefit but is also good management practice. The Forestry Commission in Britain has over 800,000 ha of forest under its management, and they provide opportunities for walking, cycling, driving, horse riding, orienteering, camping, caravaning, fishing, bird-watching and more. As many as 350 million day trips are made to the forests.

Popular though forest recreation activities might be, there are strict regulations. To enjoy the benefits of the forest, one must be mindful of the ecosystem and of other people - whether visitors or locals. Planning for forest recreational uses requires special expertise. Design of visitor uses, layout of facilities and conservation considerations - all need specific skills.

If the call for more and more recreational facilities in natural wilderness areas can be harnessed into conserving fragments of unconquered land and water resources—what many call the last remaining heritage of the past—it could contribute significantly to preserving the depleting forest cover.

Forests for Grazing Animals

Grazing forms an important forest activity. Grazing in forest areas in a planned and scientific manner has been found to be beneficial for the land. Of course, this depends on the timing and intensity. Timing is important in terms of the period allowed for regrowth between grazing. Intensity refers to how much leaf area is left on the plant after grazing.

Research shows that cattle grazing at low intensity provides biodiversity benefits to forest areas. This is because livestock grazing on ground cover reduces competition for moisture between the vegetation. Studies in Oregon have shown up to 50 per cent increase in plant growth over 10 to 20 years with the integration of animals into the system.

Though woodlands can provide useful forage for livestock, management of forest grazing resources is essential to avoid damage to plant reproduction. Many landowners perceive forests to be virtually indestructible, given the long life of trees, their ability to stand various natural stresses, and their capability of regenerating. Indiscriminate grazing is the consequence. Overgrazing can be quite counter-productive. Elimination of plant cover and compacting of soil increases soil erosion and surface run-off. Plant reproduction is affected as well, which in turn reduces forage yields.

Range Management

The science of directing forest use for maximum forage production for livestock and wildlife without damaging the forest and watershed is called range management, and it is a very specialised field. Without sensible range management, grazing can do more harm than good.

Range managers seek to increase the amount of forage on the land and reduce possible damage from livestock. Range conservation practices include: fencing the area into different pastures to allow for rotating grazing areas; development of water sources at various points for grazing animals; killing brush and weeds occupying pastures; and complete exclusion of livestock from overgrazed lands.

Forests as Watersheds

The bulk of the earth's rainfall is received by mountains, thus forming the headwaters of the land's water distribution channels. As a result, the mountain forests influence the quantity and quality of this precious freshwater resource. For example, forests are usually the best cover for safeguarding water quality from sediments and chemicals.

Flow of water from Forest catchment areas is complex, making generalisations difficult. However, there is evidence to suggest that cloud forests and some older natural forests can increase net water flow. Montane "cloud" or "fog" forests are special kinds of forests that scrape out moisture from the clouds or fog. This water extraction function is in addition to the vertical precipitation received by these forests. Having relatively low water use, these forests add to the water supply of a catchment.

Currently, about a third of the world's largest cities (33 out of 105) source a significant proportion of drinking water directly from protected forest areas. These cities include: New York, Jakarta, Tokyo, Mumbai, Rio de Janeiro, Los Angeles, Barcelona, Nairobi, and Melbourne. At least five other cities obtain water from sources that originate in distant watersheds, and at least eight more from forests that are managed to give maximum importance to providing drinking water.

Forest catchment areas can accrue significant economic value, something that goes largely unrecognised most of the time. For instance, in some circumstances, it is possible to collect user fees from stakeholders (individuals as well as companies) who benefit from drinking water in order to help pay for catchment protection benefits provided by the protected area management.

KINDS OF FORESTS

The type of forest depends mainly on location—that is, distance from equator and altitude—and climate. Broadly, forests may be classified as follows:

Tropical Rainforests

Year-round high temperatures and abundant rainfall makes this a dense, lush forest. Tropical rainforests are found near the equator. They are vital storehouses of biodiversity on the planet, and yet face severe threat today, with much of their original extent depleted.

Sub-Tropical Forests

These are found to the sound and north of the tropical forests. The trees here are adapted to resist the summer drought.

Mediterranean Forests

These forests are found to the south of the temperate regions around the coasts of the Mediterranean, California, Chile and Western Australia. The growing season is short and almost all trees are evergreen, but mixed hardwood and softwood.

Temperate Forests

Found in such places as eastern North America, northeastern Asia, and western and eastern Europe, temperate forests are a mix of deciduous and coniferous evergreen trees. Usually, the broad-leaved hardwood trees shed leaves annually. There are well-defined seasons with a distinct winter and sufficient rainfall.

Coniferous Forests

Coniferous forests inhabit the cold, windy regions around the poles. There are both hardwoods and conifers found in this region. The conifers are evergreen and structurally adapted to withstand the long drought-like conditions of the long winters, whereas the hardwoods are deciduous.

Montane Forests

These are also known as cloud forests because they receive most of their precipitation from the mist or fog that comes up from the lowlands. Some of these montane woodlands and grasslands are found in high-elevation tropical, subtropical and temperate zones. Plants and animals in these forests are adapted to withstanding the cold, wet conditions and intense sunlight. Trees are mainly conifers.

EFFECT OF DEFORESTATION

Deforestation is the conversion of forested areas to non-forest land for use such as arable land, pasture, urban use, logged area, or wasteland. Generally, the removal or destruction of significant areas of forest cover has resulted in a degraded environment with reduced biodiversity. In many countries, massive deforestation is ongoing and is shaping climate and geography.

Deforestation results from removal of trees without sufficient reforestation, and results in declines in habitat and biodiversity, wood for fuel and industrial use, and quality of life. Since about the mid-1800s the Earth has experienced an unprecedented rate of change of destruction of forests worldwide. Forests in Europe are adversely affected by acid rain and very large areas of Siberia have been harvested since the collapse of the Soviet Union. In the last two decades, Afghanistan has lost over 70 per cent of its forests throughout the country. However, it is in the world's great tropical rainforests where the destruction is most pronounced at the current time and where wholesale felling is having an adverse effect on biodiversity and contributing to the ongoing Holocene mass extinction.

About half of the mature tropical forests, between 750 to 800 million hectares of the original 1.5 to 1.6 billion hectares that once covered the planet have fallen. The forest loss is already acute in [Southeast Asia], the second of the world's great biodiversity hot spots. Much of what remains is in the [Amazon basin], where the [Amazon Rainforest] covered more than 600 million hectares.

The forests are being destroyed at an accelerating pace tracking the rapid pace of human population growth. Unless significant measures are taken on a world-wide basis to preserve them, by 2030 there will only be ten per cent remaining with another ten per cent in a degraded condition. 80 per cent will have been lost and with them the irreversible loss of hundreds of thousands of species.

Causes of Deforestation

There are many causes, ranging from slow forest degradation to sudden and catastrophic clearcutting, slash-and-burn, urban development, acid rain, and wildfires. Deforestation can be the result of the deliberate removal of forest cover for agriculture or urban development, or it can be a consequence of grazing animals, primarily for agriculture. In addition to the direct effects brought about by forest removal, indirect effects caused by edge effects and habitat fragmentation can greatly magnify the effects of deforestation.

While tropical rainforest deforestation has attracted most attention, tropical dry forests are being lost at a substantially higher rate, primarily as an outcome of slash-and-burn techniques used by shifting cultivators. Generally loss of biodiversity is highly correlated with deforestation.

Environmental Impact

Generally, the removal or destruction of significant areas of forest cover has resulted in a degraded environment with reduced biodiversity. In many countries, massive deforestation is ongoing and is shaping climate and geography. Deforestation affects the amount of water in the soil and groundwater and the moisture in the atmosphere. Forests support considerable biodiversity, providing valuable habitat for wildlife; moreover, forests foster medicinal conservation and the recharge of aquifers. With forest biotopes being a major, irreplaceable source of new drugs (like taxol), deforestation can destroy genetic variations (such as crop resistance) irretrievably.

Shrinking forest cover lessens the landscape's capacity to intercept, retain and transport precipitation. Instead of trapping precipitation, which then percolates to groundwater systems, deforested areas become sources of surface water runoff, which moves much faster than subsurface flows. That quicker transport of surface water can translate into flash flooding and more localised floods than would occur with the forest cover.

Deforestation also contributes to decreased evapotranspiration, which lessens atmospheric moisture which in some cases affects precipitation levels downwind from the deforested area, as water is not recycled to downwind forests, but is lost in runoff and returns directly to the oceans. According to one preliminary study, in deforested north and northwest China, the average annual precipitation decreased by one third between the 1950s and the 1980s.

Long-term gains can be obtained by managing forest lands sustainable to maintain both forest cover and provide a biodegradable renewable resource. Forests are also important stores of organic carbon, and forests can extract carbon dioxide and pollutants from the air, thus contributing to biosphere stability and probably relevant to the greenhouse effect. Forests are also valued for their aesthetic beauty and as a cultural resource and tourist attraction.

Economic Impact

Historically utilisation of forest products, including timber and fuel wood, have played a key role in human societies, comparable to the roles of water and cultivable land. Today, developed countries continue to utilise timber for building houses, and wood pulp for paper. In developing countries almost 3 billion people rely on wood for heating and cooking.

The forest products industry is a large part of the economy in both developed and developing countries. Short-term economic gains made by conversion of forest to agriculture, or over-exploitation of wood products, often leads to loss of long-term income. Both West Africa and Southeast Asia have experienced lower revenue because of declining timber harvests. Illegal logging causes billions of dollars of losses to national economies annually.

Characterisation of Deforestation

Throughout most of history, humans have considered forest clearing as necessary for most activities besides forestry. In most countries, only after serious shortages of wood and other forest products are policies implemented to ensure forest resources are used in a sustainable manner. Typically in developed countries, as urbanisation and economic development increases, land previously used for farming is abandoned and reverted to forests.

Today, in the developed world, most countries are experiencing forest restoration and most losses in forest land are primarily driven by expanding urban areas. In developing countries, human-caused deforestation and the degradation of forest habitat is primarily due to expansion of agriculture, slash and burn practices, urban sprawl, illegal logging, over harvest of fuel wood, mining, and petroleum exploration. It has been argued that deforestation trends follow the Kuznets curve however even if true this is problematic in so-called hot-spots because of the risk of irreversible loss of non-economic forest values for example valuable habitat or species loss.

The effects of human related deforestation can be mitigated through environmentally sustainable practices that reduce permanent destruction of forests or even act to preserve and rehabilitate disrupted forestland. These methods help the cause and provide a sustainable growth of forests and allow lumber to become a renewable resource.

Define Deforestation

Deforestation defined broadly can include not only conversion to non-forest, but also degradation that reduces forest quality—the density and structure of the trees, the ecological services supplied, the biomass of plants and animals, the species diversity and the genetic diversity. A narrow definition of deforestation is: the removal of forest cover to an extent that allows for alternative land use. The United Nations Research Institute for Social Development (UNRISD) uses a broad definition of deforestation, while the Food and Agriculture Organisation of the UN (FAO) uses a narrow definition.

Definitions can also be grouped as those which refer to changes in land cover and those which refer to changes in land use. Land cover measurements often use a per cent of cover to determine deforestation. This type of definition has the advantage in that large areas can be easily measured, for example from satellite photos. A forest cover removal of 90 per cent may still be considered forest in some cases. Under this definition areas that may have few values of a natural forest such as plantations and even urban or suburban areas may be considered forest.

Land use definitions measure deforestation by a change in land use. This definition may consider areas to be forest that are not commonly considered as such. An area can be lacking trees but still considered a forest. It may be a land designated for afforestation or an area designated administratively as forest.

It has been argued that the lack of specificity in use of the term deforestation distorts forestry issues. The term deforestation is used to refer to activities that use the forest, for example, fuel wood cutting, commercial logging, as well as activities that cause temporary removal of forest cover such as the slash and burn technique, a component of some shifting cultivation agricultural systems or clearcutting. It is also used to describe forest clearing for annual crops and forest loss from over-grazing.

Some definitions of deforestation include activities such as establishment of industrial forest plantations are considered afforestation by others. It has also been argued that the term deforestation is such an emotional term that is used "so ambiguously that it is virtually meaningless" unless it is specified what is meant.

More specific terms include forest decline, forest fragmentation and forest degradation, loss of forest cover and land use conversion. The term also has a traditional legal sense of the conversion of Royal forest land into purlieu or other non-forest land use.

The causes of deforestation are complex and often differ in each forest and country. It may be difficult to determine the cause of deforestation in a particular forest. For example, a rise in the price of soybeans may result in soybean farmers displacing cattle ranchers in order to expand their farms. This might cause cattle ranchers to shift to land previously used by slash and burn farmers. The farmers in turn shift further into the forest that has been made accessible by roads built by loggers. In this case it may not be clear who "caused" deforestation. In this case it could be claimed that while the loggers caused forest degradation and that the slash and burn farmers were agents of deforestation, the cause was demand for farm land. The underlaying causes may be poverty or the trade in international commodities.

Deforestation Theories

Three schools of thought exist with regards to the causes of deforestation: the Impoverishment school, which believes that the major cause of deforestation is "the growing number of poor," the Neoclassical school, which believes that the major cause is "open-access property rights," and the Political-ecology school which believes that the major cause of deforestation is that the "capitalist investors crowd out peasants". The Impoverishment school sees smallholders as the principal agents of deforestation, the Neoclassical school sees various agents, and the Political-ecology school sees capitalist entrepreneurs as the major agents of deforestation. Actual data support the first two theories as widespread numerical impacts.

Historical Causes of Deforestation

Deforestation has been practiced by humans since the beginnings of civilisation. Fire was the first tool that allowed humans to modify the landscape. The first evidence of deforestation shows up in the Mesolithic. It was probably used to drive game into more accessible areas. With the advent of agriculture, fire became the prime tool to clear land for crops. In Europe there is little solid evidence before 7000 BC. Mesolithic foragers used fire to create openings for red deer and wild boar.

In Great Britain shade tolerant species such as oak and ash are replaced in the pollen record by hazels, brambles, grasses and nettles. Removal of the forests led to decreased transpiration resulting in the formation of upland peat bogs. Widespread decrease in elm pollen across Europe between 8400-8300 BC and 7200-7000 BC, starting in southern Europe and gradually moving north to Great Britain, may represent land clearing by fire at the onset of Neolithic agriculture.

In ancient Greece, Tjeered van Andel and co-writers summarised three regional studies of historic erosion and alluviation and found that, wherever adequate evidence exists, a major phase of erosion follows, by about 500-1000 years the introduction of farming in the various regions of Greece, ranging from the later Neolithic to the Early Bronze Age. The thousand years following the mid-first millennium BCE saw serious, intermittent pulses of soil erosion in numerous places. The historic silting of ports along the southern coasts of Asia Minor and in coastal Syria during the last centuries BC.

The famous silting up of the harbour for Bruges, which moved port commerce to Antwerp, also follow a period of increased settlement growth (and apparently of deforestation) in the upper river basins. In early medieval Riez in upper Provence, alluvial silt from two small rivers raised the riverbeds and widened the floodplain, which slowly buried the Roman settlement in alluvium and gradually moved new construction to higher ground; concurrently the headwater valleys above Riez were being opened to pasturage.

A typical progress trap is that cities were often built in a forested area providing wood for some industry. When deforestation occurs without proper replanting, local wood supplies become difficult to obtain near enough to remain competitive, leading to the city's abandonment, as happened repeatedly in Ancient Asia Minor. The combination of mining and metallurgy often went along this self-destructive path.

Meanwhile most of the population remaining active in (or indirectly dependent on) the agricultural sector, the main pressure in most areas remained land clearing for crop and cattle farming; fortunately enough wild green was usually left standing (and partially used, e.g. to collect firewood, timber and fruits, or to graze pigs) for wildlife to remain viable, and the hunting privileges of the elite (nobility and higher clergy) often protected significant woodlands.

Major parts in the spread (and thus more durable growth) of the population were played by monastical 'pioneering' and some feudal lords actively attracting farmers to settle (and become tax payers) by offering relatively good legal and fiscal conditions—even when they did so to launch or encourage cities, there always was an agricultural belt around and even quite some within the walls. When on the other hand demography took a real blow by such causes as the Black Death or devastating warfare this could lead to settlements being abandoned, leaving land to be reclaimed by nature, even though the secondary forests usually lacked the original biodiversity.

From 1100 to 1500 AD significant deforestation took place in Western Europe as a result of the expanding human population. The large-scale building of wooden sailing ships by European (coastal) naval owners since the 15th century for exploration, colonisation, slave - and other trade on the high seas and (often related) naval warfare (the failed invasion of England by the Spanish Armada in 1559 and the battle of Lepanto 1577 are early cases of huge waste of prime timber; each of Nelson's Royal navy war ships at Trafalgar had required 6000 mature oaks) and piracy meant that whole woody regions were over-harvested, as in Spain, where this contributed to the paradoxical weakening of the domestic economy since Columbus' discovery of America made the colonial activities (plundering, mining, cattle, plantations, trade...) predominant.

In Changes in the Land, William Cronon collected 17th century New England Englishmen's reports of increased seasonal flooding during the time that the forests were initially cleared, and it was widely believed that it was linked with widespread forest clearing upstream.

The massive use of charcoal on an industrial scale in Early Modern Europe was a new acceleration of the onslaught on western forests; even in Stuart England, the relatively primitive production of charcoal has already reached an impressive level. For ship timbers, Stuart England was so widely deforested that it depended on the Baltic trade and looked to the untapped forests of New England to supply the need. In France, Colbert planted oak forests to supply the French navy in the future; as it turned out, as the oak plantations matured in the mid-nineteenth century, the masts were no longer required.

Norman F. Cantor's summary of the effects of late medieval deforestation applies equally well to Early Modern Europe:

"Europeans had lived in the midst of vast forests throughout the earlier medieval centuries. After 1250 they became so skilled at deforestation that by 1500 AD they were running short of wood for heating and cooking. They were faced with a nutritional decline because of the elimination of the generous supply of wild game that had inhabited the now-disappearing forests, which throughout medieval times had provided the staple of their carnivorous high-protein diet. By 1500 Europe was on the edge of a fuel and nutritional disaster, [from] which it was saved in the sixteenth century only by the burning of soft coal and the cultivation of potatoes and maize."

Specific parallels are seen in twentieth century deforestation occurring in many developing nations.

Environmental Effects of Deforestation

Atmospheric Pollution

Deforestation is often cited as one of the major causes of the enhanced greenhouse effect. Trees and other plants remove carbon (in the form of carbon dioxide) from the atmosphere during the process of photosynthesis. Both the decay and burning of wood releases much of this stored carbon back to the atmosphere. A. J. Yeomans asserts in Priority One that overnight, as trees consume the sugars that they produced during the day, a stable forest releases exactly the same quantity of carbon dioxide back into the atmosphere. Others state that mature forests are net sinks of CO_2. Deforestation caused by humans is estimated to contribute to one-third of all carbon dioxide.

The water cycle is also affected by deforestation. Trees extract groundwater through their roots and release it into the atmosphere. When part of a forest is removed, the region can not hold as much water and can result in a much drier climate.

Biological Diversity

Some forests are rich in biological diversity. Deforestation can cause the destruction of the habitats that support this biological diversity, thus causing contributing to the ongoing Holocene extinction event. Numerous countries have developed Biodiversity Action Plans to limit clearcutting and slash and burn agricultural practices as deleterious to wildlife and vegetation, particularly when endangered species are present.

The diverse species within rainforests has long been a useful area for research and learning. Apes and other primates in their natural environment are a source of notable research. Numerous significant medications have been developed from genetic materials within forests, many of which pertain to endangered species. Deforestation can subject some of these genetic materials to irreversible loss.

Hydrologic Cycle

Trees, and plants in general, affect the hydrological cycle in a number of significant ways:

1. Their canopies intercept a proportion of precipitation, which is then evaporated back to the atmosphere (canopy interception);
2. Their litter, stems and trunks slow down surface runoff;
3. Their roots create macropores—large conduits—in the soil that increase infiltration of water;
4. They contribute to terrestrial evaporation and reduce soil moisture via transpiration;

5. Their itter and other organic residue change soil properties that affect the capacity of soil to store water.

As a result, the presence or absence of trees can change the quantity of water on the surface, in the soil or groundwater, or in the atmosphere. This in turn changes erosion rates and the availability of water for either ecosystem functions or human services. The forest may have little impact on flooding in the case of large rainfall events, which overwhelm the storage capacity of forest soil if the soils are at or close to saturation.

Soil Erosion

Undisturbed forest has very low rates of soil loss, approximately 0.02 metric tons per hectare. Deforestation generally increases rates of soil erosion, by increasing the amount of runoff and reducing the protection of the soil from tree litter. This can be an advantage in excessively leached tropical rain forest soils. Forestry operations themselves also increase erosion through the development of roads and the use of mechanised equipment.

China's Loess Plateau was cleared of forest millennia ago. Since then it has been eroding, creating dramatic incised valleys, and providing the sediment that gives the Yellow River its yellow colour and that causes the flooding of the river in the lower reaches (hence the river's nickname 'China's sorrow').

Removal of trees does not always increase erosion rates. In certain regions of southwest US, shrubs and trees have been encroaching on grassland. The trees themselves enhance the loss of grass between tree canopies. The bare intercanopy areas become highly erodible. The US Forest Service, in Bandelier National Monument for example, is studying how to restore the former ecosystem, and reduce erosion, by removing the trees.

Tree roots bind soil together, and if the soil is sufficiently shallow they act to keep the soil in place by also binding with underlying bedrock. Tree removal on steep slopes with shallow soil thus increases the risk of landslides, which can threaten people living nearby. However most deforestation does not affect the trunks of tree's allowing for the roots to stay rooted negating the landslide.

Deforestation Control

Intensive Farming

New methods are being developed to farm more intensively, such as high-yield hybrid crops, greenhouse, autonomous building gardens, and hydroponics. These methods are often dependent on massive chemical inputs to maintain necessary yields. In cyclic agriculture, cattle are grazed on farm land that is resting and rejuvenating. Cyclic agriculture actually increases the fertility of the soil. Intensive farming can also decrease soil nutrients by consuming at an accelerated rate the trace minerals needed for crop growth.

Forest Management

Efforts to stop or slow deforestation have been attempted for many centuries because it has long been known that deforestation can cause environmental damage sufficient in some cases to cause societies to collapse. In Tonga, paramount rulers developed policies designed to prevent conflicts between short-term gains from converting forest to farmland and long-term problems forest loss would cause, while during the seventeenth and eighteenth centuries in Tokugawa Japan the shoguns developed a highly sophisticated system of long-term planning to stop and even reverse deforestation of the preceding centuries through substituting timber by other products and more efficient use of land that had been farmed for many centuries.

In sixteenth century Germany landowners also developed silviculture to deal with the problem of deforestation. However, these policies tend to be limited to environments with good rainfall, no dry season and very young soils (through volcanism or glaciation). This is because on older and less fertile soils trees grow too slowly for silviculture to be economic, whilst in areas with a strong dry season there is always a risk of forest fires destroying a tree crop before it matures.

Reforestation

In the People's Republic of China, where large scale destruction of forests has occurred, the government has in the past required that every able-bodied citizen between the ages of 11 and 60 plant three to five trees per year or do the equivalent amount of work in other forest services. The government claims that at least 1 billion trees have been planted in China every year since 1982. This is no longer required today, but March 12 of every year in China is the Planting Holiday.

In western countries, increasing consumer demand for wood products that have been produced and harvested in a sustainable manner are causing forest landowners and forest industries to become increasingly accountable for their forest management and timber harvesting practices.

The Arbor Day Foundation's Rain Forest Rescue programme is a charity that helps to prevent deforestation. The charity uses donated money to buy up and preserve rainforest land before the lumber companies can buy it.

The Arbor Day Foundation then protects the land from deforestation. This also locks in the way of life of the primitive tribes living on the forest land. Organisations such as The Nature Conservancy, World Wide Fund for Nature, Conservation International, African Conservation Foundation and Greenpeace also focus on preserving forest habitats.

Forest Plantations

To meet the worlds demand for wood it has been suggested by forestry writers Botkins and Sedjo that high-yielding forest plantations are suitable. It has been calculated that plantations yielding 10 cubic metres per hectare annually could supply all the timber required for international trade on 5 per cent of the world's existing forestland. By contrast natural forests produce about 1-2 cubic metres per hectare, therefore 5 to 10 times more forest land would be required to meet demand. Forester Chad Oliver has suggested a forest mosaic with high-yield forest lands interpersed with conservation land.

The Jewish National Fund states that the only country to come out of the Twentieth Century with more trees than it had at the start of the period was Israel.

8

ENERGY AND ENVIRONMENT

The importance of energy supply systems, both industrial and traditional, in the mobilisation of such toxic substances as sulphur oxides and particles as well as in the release of carbon dioxide, the principal greenhouse gas. Also shown is the human disruption index for each substance, which is the ratio of the amount released by human activities to natural releases. This indicates that together with other human activities, energy systems are significantly affecting the cycling of important chemical species at the global scale. Although by themselves these indexes do not demonstrate that these insults are translated into negative impacts, their magnitudes provide warning that such impacts could be considerable.

In the past hundred years most of these phenomena have grown from local perturbations to global disruptions. The environmental transition of the 20th century-driven by more than 20-fold growth in the use of fossil fuels and augmented by a tripling in the use of traditional energy forms such as biomass-has amounted to no less than the emergence of civilisation as a global ecological and geochemical force.

The impacts from energy systems, however, occur from the household to the global scale. Indeed, at every scale the environmental impacts of human energy production and use account for a significant portion of human impacts on the environment.

HOUSEHOLD FUELS

The oldest human energy technology, the home cooking fire, persists as the most prevalent fuel-using technology in the world. For much of the world's population, the home cooking fire accounts for most direct energy demand. Household fuel demand accounts for more than half of energy demand in most countries with per capita incomes under $1,000.

The 'energy ladder' is a useful framework for examining trends and impacts of household fuel use. The ladder ranks household fuels along a spectrum running from simple biomass fuels (dung, crop residues, wood) through fossil fuels (kerosene and gas) to the most modern form (electricity). The fuel-stove combinations that represent rungs in the ladder tend to become cleaner, more efficient, more storable, and more controllable in moving up the ladder. But capital costs and dependence on centralised fuel cycles also tend to increase in moving up the ladder.

Although there are local exceptions, history has generally shown that when alternatives are affordable and available, population tend to move up the ladder to higher-quality fuel-stove combinations. Although all of humanity had its start a quarter of a million years ago at the top of the energy ladder in those times (wood), only about half has moved up to higher-quality rungs. The remaining half is either still using wood or has been forced by local wood shortages down the ladder to crop residues, dung, or, in some severe situations, to the poorest-quality fuels such as shrubs and grass.

Throughout history in places where coal is easily available, local wood shortages have led some population to move to coal for household use. This shift occurred about a thousand years ago in the United Kingdom, for example, although it is relatively uncommon there today. In the past 150 years such transitions occurred in Eastern Europe, China and India, where coal use still persists in millions of households. In terms of the energy ladder, coal represents an upward movement in terms of efficiency and storability. Because of these characteristics and its higher energy densities, it is possible to ship coal economically over longer distances than wood and to efficiently supply urban markets. In these senses, coal is like other household fossil fuels. Unlike kerosene and gas, however, coal is often a dirtier fuel than wood.

Harvesting and Combustion

The chief environmental impacts of household fuel cycles relate to harvesting and combustion. In the 1970s books and newspapers called attention to the 'other' energy crisis, referring to the growing and alarming shortage of woodfuel affecting a large fraction of the world population that depended on it. Since deforestation and desertification were often also occurring in such places, it was perhaps a logical conclusion that fuel demand was to blame. It is still common today to read that deforestation is caused by fuel gathering in rural areas of developing countries The most important cause by far seems to be expansion of agricultural lands, followed by lumbering and road building. Indeed, the causation is more often the reverse-that is, the shortage of fuelwood is due to deforestation, rather than the other way around.

Part of the misunderstanding stems from the assumption that rural households gather woodfuel from forests. In many areas villagers gather significant amounts of fuelwood from what Gerald Leach has called 'invisible trees'—the trees around fields, next to houses, along roads, and so on that do not show up in most satellite remote sensing surveys or national forest statistics. Thus when estimates of local fuelwood demand appear to exceed growth rates in local forests, it does not necessarily imply that deforestation is taking place. Conversely, if deforestation is known to be occurring, it does not mean that fuel demand is necessarily the reason.

Similarly, desertification in the Sahel and elsewhere in Sub- Saharan Africa has links to fuel demand. But it has been difficult to separate out the influence of all the relevant factors, including climate change, intensification of grazing, land-use shifts, and fuel harvesting. Nevertheless, as with deforestation, there are some poor areas where harvesting of wood and brush plays an important role. Although the link between fuelwood harvesting and deforestation is far from universal, there are localised cases in which fuelwood demand seems to contribute significantly to forest depletion. Most prominent among these are places, mainly in Sub-Saharan Africa, where commercial charcoal production is practised. In these areas temporary kilns (legal or illegal) in forested areas are used until local wood resources are depleted, then moved or rebuilt elsewhere.

Charcoal, being a relatively high-quality and high-density fuel, can be trucked economically across long distances to urban markets. Thus large charcoal-using cities can have 'wood sheds' extending hundreds of kilometres along roadways, though there is evidence that significant regrowth often occurs over long enough periods. In some arid and semiarid areas, harvesting by woodfuel traders to meet urban demand seems to contribute to forest depletion, although, again, regrowth is often occurring. The quality of regrowth in terms of biodiversity and other ecosystem parameters is not well documented, however.

The harvesting of dung and crop residues as fuel does not have much direct environmental impact. But in some areas it may deprive local soils of needed nutrients and other conditioners. Indeed, in most rural areas the use of dung as fuel rather than fertiliser is probably a sign of poverty and lack of fuel alternatives. Crop residues, on the other hand, consist of a wide variety of materials, many of which do not have much value as fertiliser or soil conditioner. Indeed, in some cases disposal becomes a serious problem if these residues are not gathered for fuel. In these cases the usual practice is to burn the residues in place on the fields, with consequent pollution implications (although sometimes with benefits in terms of pest control).

Consequently, regardless of development level, air pollution from post-harvest burning of farmland is a significant seasonal source of air pollution in many agricultural areas around the world. Harvesting and preparation of household biomass fuels also have occupational health impacts on women and children, due, for example, to heavy loads and burns.

Combustion

Physical form and contaminant content are the two characteristics of fuels that most affect their pollutant emissions when burned. It is generally difficult to pre-mix solid fuels sufficiently with air to assure good combustion in simple small-scale devices such as household stoves. Consequently, even though most biomass fuels contain few noxious contaminants, they are usually burned incompletely in household stoves and so produce a wide range of health-damaging pollutants. Wood and other biomass fuels would produce little other than non-toxic products, carbon dioxide, and water when combusted completely. But in practice sometimes as much as one-fifth of the fuel carbon is diverted to products of incomplete combustion, many of which are important health-damaging pollutants. Coal, on the other hand, is not only difficult to burn completely because it is solid, but also often contains significant intrinsic contaminants that add to its emissions of health-damaging pollutants.

Most prominent among such emissions are sulphur oxides. But in many areas coal also contains arsenic, fluorine, and other toxic elements that can lead to serious health-damaging pollutants. Tens of millions of people in China, for example, are exposed to such pollutants from household coal use.

Petroleum-based liquid and gaseous fuels, such as kerosene and liquefied petroleum gas, can also contain sulphur and other contaminants, though in much smaller amounts than in many coals. In addition, their physical forms allow much better pre-mixing with air in simple devices, assuring much higher combustion efficiencies and lower emissions of health-damaging pollutants in the form of products of incomplete combustion. Furthermore, stoves for petroleumbased liquid and gaseous fuels are much more energy efficient than those for coal. As a result emissions of health-damaging pollutants per meal from these fuels are at least an order of magnitude less than those from solid fuels.

Not only do solid-fuel stoves produce substantial emissions of health-damaging pollutants per meal, but a large fraction do not have chimneys for removing the emissions from the home. Consequently, indoor concentrations of health-damaging pollutants can reach very high levels.

The energy flows of a typical wood-fired cooking stove, in which a large fraction of the fuel energy is lost because of low combustion efficiency or low transfer of the heat to the pot. The excessive pollutant levels commonly reached in these circumstances, well beyond World Health Organisation guidelines. Even in households with chimneys, however, heavily polluting solid-fuel stoves can produce significant local outdoor pollution. This is particularly true in dense urban slums, where such 'neighbourhood' pollution can be much higher than average urban pollution levels.

To estimate the health damage from pollution, it is necessary to take into account the amount of pollution released. Equally important, however, is the behaviour of the population at risk. Even a large amount of pollution will not have much health impact if little of it reaches people. But a relatively small amount of pollution can have a big health impact if it is released at the times and places where people are present, such that a large fraction is breathed in. Thus it is necessary to look not only at where the pollution is but also at where the people are.

Unfortunately, pollution from household stoves is released right at the times and places where people are present-that is, in every household every day. This is the formula for high pollution exposures: significant amounts of pollution often released in poorly ventilated spaces at just the times when people are present. Moreover, because of their nearly universal responsibility for cooking, women and their youngest children are generally the most exposed.

Health Effects of Air Pollution

Considering the sizes of the relevant population and the exposures to health-damaging pollutants, there has been relatively little scientific investigation of the health effects of indoor air pollution in developing countries relative to studies of outdoor air pollution in cities. Nevertheless, enough has been done to enable rough estimates of the total impact of air pollution, at least for women and young children (who suffer the highest exposures).

Four main types of health effects are thought to occur, based on studies in households that use solid fuels and corroborated by studies of active and passive smoking and outdoor air pollution:

1. Infectious respiratory diseases such as acute respiratory infections and tuberculosis.
2. Chronic respiratory diseases such as chronic bronchitis and lung cancer.
3. Adverse pregnancy outcomes such as stillbirth and low birthweight in babies born to women exposed during pregnancy.
4. Blindness, asthma, and heart disease (less evidence to date).

The best estimates of such effects for developing countries have been done for India. These indicate that household solid fuel use causes about 500,000 premature deaths a year in women and children under 5. This is 5-6 per cent of the national burden of ill health, or 6-9 per cent of the burden for these two population groups. This is comparable to, though somewhat less than, the estimated national health impacts of poor water and sanitation at the household level-and more than the national burdens of such major health hazards as malaria, tuberculosis, tobacco, AIDS, heart disease, or cancer.

Given that India contains about one-quarter of the world's solid-fuel cooking stoves, the global impact could be expected to be about four times larger, or about 2 million deaths a year in women and children. This is roughly compatible with World Health Organisation estimates of about 2.5 million, estimates that were generated by extrapolating studies from industrialised country cities to developing country conditions.

Greenhouse Gas Emissions

The same incomplete combustion processes that produce emissions of health-damaging pollutants from household solid-fuel stoves also produce greenhouse gas emissions. A large amount of fuel carbon is typically diverted to gaseous products of incomplete combustion, all of which cause greater global warming per carbon atom than would be the case if complete combustion occurred and all the carbon was released as carbon dioxide. The most powerful of these is methane, which over a 20-year period causes more than 20 times the global warming from the same amount of carbon as carbon dioxide.

Greenhouse gas emissions from several of the most important household fuels in developing countries. Because of significant emissions of non-carbon dioxide greenhouse gases, solid biomass fuels, even though renewable, can have a larger greenhouse gas commitment per meal than fossil fuels, kerosene, and liquefied petroleum gas. These relationships have several important policy implications:

1. Even if renewably harvested, many biomass fuel cycles are not greenhouse gas neutral because of their substantial production of products of incomplete combustion.
2. In some situations, therefore, substitution of fossil fuels for renewable biomass might be recommended to reduce greenhouse gas emissions.
3. To be greenhouse gas neutral, biomass fuel cycles must be based on renewable harvesting and must have close to 100 per cent combustion efficiency (which most do not in their current configurations).
4. Improved biomass stoves should be designed to increase overall efficiency and to reduce combustion inefficiency, which is the cause of greenhouse gas and health-damaging pollutants.

Stoves using biogas, which is made in household or village anaerobic digesters from dung, have by far the least greenhouse gas emissions per meal-only about 10 per cent of those for liquefied petroleum gas and a factor of 80 less than the average stove burning dung directly.

A complete comparison of these fuel-stove combinations would require evaluating greenhouse gas emissions over the entire fuel cycle in each case, for example, including methane leaking from biogas digesters and releases from oil refineries making kerosene. Nevertheless, the extremely low greenhouse gas emissions from biogas stoves illustrate the potentially great advantage for greenhouse gas emissions of processed biomass fuels such as biogas. Such fuels can be both renewably harvested as well as burned as liquids or gases with high combustion efficiency. Reducing the human health and global warming impacts of household stoves will require better stoves with higher efficiency, lower emissions, and cleaner fuels.

WORK AND WORKPLACES

The extraction, transport, use, and waste management of energy sources involve important health hazards related to the work and workplaces involved in these activities. Many of the jobs involved, such as forestry and mining for solid fuels, are particularly dangerous. Many workers are engaged in these jobs, particularly in countries that are rapidly developing their industries and the energy sources that the industries require.

Role of Biomass

Wood, crop residues, dung, and the like are common energy sources for poor households in developing countries. Wood is also still widely used in industrialised countries, in some cases promoted in the interest of reducing greenhouse gas emissions. Wood and agricultural waste are often collected by women and children. Such collection is part of daily survival activities, which also include water hauling, food processing, and cooking. An analysis in four developing countries found that women spend 9-12 hours a week on these activities, while men spend 5-8 hours. Women's role in firewood collection is most prominent in Nepal (2.4 hours a day for women and 0.8 hours for men).

Firewood collection may be combined with harvesting of wood for local use in construction and small-scale cottage industry manufacturing. This subsistence work is often seasonal, unpaid, and not recorded in national economic accounts. Globally about 16 million people are involved in forestry, more than 14 million of them in developing countries and 12.8 million in subsistence forestry. A number of health hazards are associated with the basic conditions of the forest. Forest workers have a high risk of insect bites, stings from poisonous plants, leech infestation, cuts, falls, and drowning. In tropical countries the heat and humidity put great strain on the body, while in temperate countries the effects of cold are a potential hazard.

The work is outside, and in sunny countries ultraviolet radiation can be another health hazard, increasing the risk of skin cancer and eye cataracts. All forestry work is hard physical labour, with a risk of ergonomic damage such as painful backs and joints as well as fatigue, which increases the risk of injuries. Heavy loads of firewood contribute to ergonomic damage. Women carrying heavy loads of firewood are a common sight in areas with subsistence forestry. Falling trees, sharp tools, dangerous machinery, and falls from heights are the main causes of injuries.

Urbanisation leads to the development of a commercial market for firewood and larger-scale production of firewood from logs or from smaller waste material left over after logs have been harvested. Energy forestry then becomes more mechanised, exposing workers to additional hazards.

Motorised hand tools (such as chain saws) become more common, resulting in high risks of injuries, noise-induced hearing losses, and 'white finger disease' caused by vibration of the hands. In addition, fertilisers and pesticides become part of the production system, with the potential for poisoning those who spray pesticides. As forestry develops further, more logging becomes mechanised, with very large machinery reducing the direct contact between workers and materials. Workers in highly mechanised forestry have only 15 per cent of the injury risk of highly skilled forestry workers using chain saws. Still, firewood production remains an operation that requires manual handling of the product at some stage and so tends to remain hazardous.

Role of Coal

Coal is a major global energy source, accounting for 23 per cent of total energy consumption. It was the primary energy source from 1900 until 1960, when it was overtaken by oil. Coal can be produced through surface (open cast) mining or underground mining.

Like mining in general, both operations are inherently dangerous to the health of the workers. About 1 per cent of the global workforce is engaged in mining, but these workers account for 8 per cent of the 15,000 fatal occupational accidents each year. Armstrong and Menon offer a detailed review of occupational health and safety issues in coal mining and other mining.

Underground coal miners are exposed to the hazards of excavating and transporting materials underground. These hazards include injuries from falling rocks and falls into mine shafts, as well as injuries from machinery used in the mine. There are no reliable data on injuries of this type from developing countries, but in industrialised countries miners have some of the highest rates of compensation for injuries-and the situation is likely to be worse in developing countries.

Other health hazards specific to underground coal mining include coal dust, which can cause 'coal worker's pneumoconiosis' or anthracosis, often combined with silicosis. Coal dust is explosive, and explosions in underground coal mines are constant danger for coal miners. Coal inherently burns, and fires in coal mines are not uncommon. Once such fire has started, it may be almost impossible to extinguish. Apart from the danger of burns, the production of smoke and toxic fumes create great health risks for the miners.

Even without fires, the coal material produces toxic gases when is disturbed: carbon monoxide, carbon dioxide, and methane. Carbon monoxide is extremely toxic because it binds to haemoglobin in the blood, blocking oxygen transport and creating chemical suffocation. Carbon monoxide is a colourless and odourless gas and so gives no warning before the symptoms of drowsiness, dizziness, headache, and unconsciousness occur. Carbon dioxide, also colourless and odourless, displaces oxygen in underground air and can also cause suffocation. Another health hazard in mining is exhaust from the diesel engines used in underground machinery and transport vehicles. This exhaust contains very fine particles, nitrogen oxides, and carbon monoxide, all of which pose serious health hazards. Exposure to fine particles in diesel exhaust increases the risk of lung cancer.

Surface coal mining avoids some of the hazards of working underground. Still, it involves the risk of injuries from machinery, falls, and falling rocks. In addition, coal mining is energy-intensive work, and heat, humidity, and other weather factors can affect workers' health. The machinery used is also noisy, and hearing loss is a common among miners. Another health hazard is the often squalid conditions under which many coal workers and their families in developing countries live, creating particular risk for the diseases of poverty.

After extraction, coal needs to be processed and transported to residential areas, power stations, and industries. This creates other types of occupational hazards. For instance, coal for residential use is often ground and formed into briquettes. This work involves high levels of coal dust as well as noise hazards. Loading, transportation, and off-loading of large amounts of coal involves ergonomic, noise, and injury hazards.

The large-scale use of coal in power stations and industry creates yet more hazards. One is the conversion of coal to coke in steel production. This process distils a large number of volatile polycyclic aromatic hydrocarbons in coal, the so-called coal tar pitch volatiles. Exposure to these hydrocarbons puts coke oven workers at twice the lung cancer risk of the general population.

Oil and Gas Exploration

Oil and gas exploration, drilling, extraction, processing, and transport involve a number of the same hazards as mining: heavy workload, ergonomic hazards, injury risk noise, vibration, and chemical exposures. This work is often carried out in isolated areas with inclement weather conditions. Long-distance commuting may also be involved, causing fatigue, stress, and traffic accident risks. The ergonomic hazards lead to risks of back pain and joint pain. Injury hazards include burns and explosions. Skin damage from exposure to oil and to chemicals used in drilling creates a need for well-designed protective clothing. A lot of exploration and drilling for oil and gas occur offshore. This involves underwater diving, which is dangerous.

Hydroelectric Power

Major hazards occur when a hydroelectric power station is built, because this usually requires constructing a large dam, excavating underground water channels, and building large structures to house the generator. McManus lists 28 occupational hazards potentially involved in the construction and operation of hydroelectric power stations. These include asbestos exposure, diesel and welding fumes, work in confined spaces or awkward positions, drowning, electrocution, noise, heat, electromagnetic fields, vibration, weather-related problems, and chemical exposures from paints, oils, and PCBs (polychlorinated biphenyls). As in any industry, however, proper attention to health and safety can keep the risks to acceptable levels. The manufacture of wind and solar power equipment involves the typical hazards in manufacturing: injuries, noise, chemical exposures, and so on.

Nuclear Power

Nuclear power generation has its own hazards due to the radiation danger involved in mining, processing, and transporting uranium, as well as the radiation in nuclear power stations. In addition, occupational hazards will develop as countries start to deal with the backlog of radioactive waste. Due to the major potential risk to the general public from a malfunctioning nuclear power station, the safety of stations is always paramount. This has contributed to a low average occupational health risk for workers in the stations.

The mining of uranium has been an important occupational health hazard in nuclear power generation, as underground mining for uranium often entails high exposure to radon, a radioactive gas emitted from uranium. Radon exposure leads to an increased risk of lung cancer.

Until the Chernobyl accident, relatively few nuclear power station workers had been affected by radiation exposure. In that accident, however, 40 workers lost their lives in the fire or due to acute radiation exposure. The long-term impact on workers exposed during the accident in the form of cancer and other radiation related effects is not yet known, however. The clean-up after the accident may eventually create substantially more effects. As many as 900,000 army, police, and other workers were called on to take part. Many workers were needed because they were only allowed to work for a short time, until they had reached the maximum allowable radiation dose. In some cases this dose was reached in a few minutes.

Workers and Quantitative Health Effects

It is difficult to estimate the number of workers involved in meeting the energy requirements of communities. As noted, in poor communities much of this work is carried out by family members, particularly women, who are not formally employed. An estimated 16 million people are involved in forestry, most of them in developing countries. In industrialised countries with reliable statistics, the occupational mortality rate for agricultural workers is 5-10 times the average for all workers. Because of the additional risks in forestry, mortality rates for these workers are possibly twice as high again, or 32,000-160,000 at a global level. Not all of this activity is directly related to fuel demand, however.

Miners are a large occupational group in international statistics. They represent up to 2 per cent of the economically active population in certain developing countries. Mining is an extremely dangerous occupation. Recent data show that occupational mortality rates for miners are up to 20 times the average for all occupations. The range of mortality rates may be as wide as that for forestry (2-10 per 1,000 workers per year).

In most countries the economically active population is 40-60 per cent of the population over 15. Thus miners may account for 1 per cent of the population over 15, or about 30 million people world-wide. If half of these miners are coal miners, the number of miners killed each year in accidents would be about the same as for forestry workers (30,000-150,000).

Another approach to this calculation is through total coal production. If applied to the world's coal production today, about 70 per cent of which is in developing countries, the mean death rate in U.S. mines from 1890-1939 of 3.1 deaths per million tons produced would predict 16,000 coal mining deaths a year world-wide. This may be low, however, because China alone has about 6,500 coal mining deaths a year according to official statistics, which tend to be incomplete. The estimate of 6,500 of 16,000 deaths, on the other hand, is roughly consistent with China's 30 per cent share of global production.

For energy production and distribution as a whole, occupational mortality may sum to 70,000-350,000 a year. These numbers are likely to exclude many cases of occupational disease (such as cancers caused by asbestos or radiation) and deaths among the many workers in informal workplaces. The upper limit of the numbers, however, may also be inflated by the crude estimates of mortality rates and number of workers. Occupational mortality rates in energy jobs in industrialised countries are generally 10-30 times lower than in developing countries, indicating that more effective prevention programmes could eliminate more than 90 per cent of the deaths referred to above. Still, energy-related jobs have inherent health risks that need to be considered when assessing the full impact of energy production and distribution.

COMMUNITY LEVEL

Energy systems are associated with a vast array of insults and impacts. Many of these are expressed at the community scale, including problems associated with coal and uranium mining, petroleum and gas extraction, water use and contamination by power plants, thermal pollution, and noise from wind farms. Here we can only focus on the largest of these impacts world-wide.

Urban air pollution is the chief environmental impact of energy systems at the community level. Although there are industrial and other sources of some pollutants, the vast bulk-whether measured by mass or by hazard-is generated by fuel combustion or, as in the case of photochemical smog, is created in the urban atmosphere by precursor chemicals largely released in the course of fuel use. From the 1930s to the 1950s a number of urban air pollution episodes in the industrialised world brought air pollution to the attention of the public. The first major improvements came by banning the burning of refuse and coal within city limits. By the early 1970s the infamous London smogs (and their parallels in other cities), caused by coal combustion, were memories.

Fuel-derived Air Pollution in Industrialised World

During the past 25 years the cities of the industrialised world have generally brought energy-derived urban air pollutants under even greater control. In the United States, for example, emissions per unit of useful energy from power plants and automobiles-the two largest urban energy polluters-have fallen 65 per cent and 50 per cent in health hazard (weighted by the relative standards for each pollutant). Japan and Western Europe have achieved similar results.

In the power sector these achievements have mostly come about by relying more on nuclear power and natural gas and by requiring smokestack controls for particles and nitrogen and sulphur oxides at coal-fired power plants. In addition, thermal power plants have become more efficient, and more improvements are expected, particularly for those using gas. For vehicles, the reductions have come from a mix of improvements in engine combustion, increases in fuel efficiency (in North America), and the nearly universal requirement of catalytic converters (devices to help control pollutant emissions). Thus, despite significant increases in power production and vehicle use since 1975, overall emissions of most pollutants are now lower.

As a result of these emission reductions, urban air quality has generally improved throughout the industrialised world. Although fuel combustion produces a number of health-damaging pollutants, as explained above, small particles are probably the best single indicator. Suspended small particles are a mix of primary combustion particles-carbonaceous materials and associated trace elements-and secondary conversion products such as sulphate and nitrate aerosols. In many parts of the world, windblown and urban dust can also be significant contributors to suspended particles.

Small particles are deposited deep in the lungs, where their clearance is slow and their ability to cause damage is enhanced. Small particles also carry adsorbed trace metals and carcinogenic hydrocarbons into the lungs, intensifying the potential for health damage. Assessments of the human health effects of air pollutants increasingly focus on these small particles. Still, there are few measurements of these particles in most cities, although more cities are measuring PM10 (particles less than 10 micrograms), which is considered a better indicator than simple total particulate levels.

In the late 1990s the mean annual cocentration of PM10 in North American, Western European, and

Japanese cities ranged from 30-45 micrograms per cubic metre. In the 1960s particulate levels were probably two to four times higher. (Small particles were not measured routinely until the mid-1980s, so previous levels have to be inferred from measurements of total particles.) Still, industrialised countries face a number of energy-related air pollution challenges. Nitrogen dioxide and ozone levels exceed standards in many cities, particularly in sunny cities with large auto fleets such as Los Angeles (California) and Athens (Greece).

The recent evidence suggesting that small particles (less than 2.5 micrograms) may be even better indicators of ill health than PM10 has led the United States to propose new regulations aimed at PM2.5, potentially putting a number of cities out of compliance. European countries are also considering such regulations. Since long-term data are not widely available, it is not clear how much PM2.5 levels have decreased in recent decades, partly because such particles are transported over much larger areas than larger particles.

This focus on even smaller particles has brought diesel exhaust particles under more scrutiny. Unlike gasoline, diesel produces a significant amount of emissions of particles that are not only smaller but may have chemical properties that make them more dangerous. This feature raises questions about the future of diesel-fuelled vehicles, even though such vehicles can be slightly more fuel-efficient and cost-effective than gasoline-fuelled vehicles. The tendency for many countries to keep diesel prices low relative to gasoline-as a means of assisting farming, fishing, and other industries-can artificially promote diesel passenger vehicles.

Since the 1980s studies have seemed to indicate that there is no threshold for the health effects of particle pollution. In other words, there no longer seems to be an identifiable level that can be termed safe. All that can be said is that the effect is lower at lower levels, but does not seem to disappear even at the lowest (background) concentrations. Indeed, in the late 1990s European and global offices of the World Health Organisation revised their particle guidelines to reflect the absence of thresholds.

Because it is rarely (if ever) practical to set a standard of zero for pollutants with significant natural sources, standard setting is much harder for pollutants with no threshold for significant effects. Policy-makers must determine that the benefits of fuel combustion outweigh the extra mortality produced by the resulting pollution-for example, that the 5 per cent increase in mortality 'allowed' by a PM10 standard of 50 micrograms per cubic metre above background is acceptable given the societal advantages of fuel use. This is a difficult determination, and much more politically difficult than endorsing a standard that has some scientific validity of being below a 'no effects' level, which is how most standards are set. The likely result will be continuous pressure to tighten particle standards, with stronger incentives for lower particle emissions from vehicles, power plants, and other fuel-using sources.

Air Pollution in Cities of Developing Countries

Developing country cities have much higher mean pollutant concentrations than industrialised country cities. In cities in China and India, averages seem to be nearly 200 micrograms per cubic metre of PM10, though there is much variation by season and city. Such concentrations must be causing significant premature deaths-perhaps 15 per cent or more above background levels. Indeed, estimates for premature mortality from urban air pollution range from 170,000-290,000 a year in China and 90,000-200,000 in India.

The causes of air pollution in developing country cities are much more varied than in industrialised countries. Although automobile ownership rates are much lower, there tend to be many other types of vehicles as well, including two- and three-wheelers using highly polluting two-stoke engines. There also tend to be larger portions of light-duty and heavy-duty vehicles using diesel rather than gasoline. In addition to power plants and large industries with limited pollution controls, developing country cities tend to have large numbers of small boilers, engines, and open fires in the commercial and light-industry sectors, as well as in informal sectors such as street food. These enterprises tend to rely on the most inexpensive and thus dirty fuels in inefficient applications without pollution controls-and so have high emissions per unit of useful energy.

Furthermore, such cities often do not have adequate refuse and garbage collection, leading to street-side trash burning, a highly polluting activity. Even when collected, trash often burns uncontrollably in landfills in or near cities, wasting potential energy and producing clouds of noxious emissions. Another major non-energy source of particle pollution in many cities is dust from construction sites and unmanaged empty land. Finally, unlike in industrialised countries, a large fraction of urban households in developing countries still use solid fuels for cooking and space heating in inefficient stoves with no pollution controls. Although individually small, their large number means that these stoves can contribute significantly to urban pollution.

Urban Pollution Control

Because the best commercial technology in terms of energy efficiency and emissions has not been deployed completely in industrialised countries and has been used little in developing countries, much improvement is possible in the next 20 years without switching to advanced technologies. In the longer term, however, if air pollution levels are to be brought down to and kept at low levels given the projected increase in population, urbanisation, economic activity, and energy use, it will be necessary to develop and deploy new, even cleaner and more efficient energy-using technologies. A number of advanced power plant technologies potentially offer such performance.

Hydroelectric dams, large and small, have brought tremendous benefits to many regions, including important contributions to development in industrialised countries. It is important not to deny these benefits to developing countries. But such dams need to be designed and constructed with care.

Although dams frequently serve many purposes-including flood control, irrigation, navigation, and recreation-major dams (those over 150 metres high, with 25 cubic kilometres of storage, or 1,000 megawatts of electricity) tend to have hydropower as one of the their main objectives. Such dams often have big impacts on the environment. There are more than 300 major dams world-wide, and nearly all have hydropower as a major component of their function. The environmental impact per unit of electricity production, however, can often be smaller for large than for small dams. The type rather than the size can be the most important factor.

With a total capacity of about 640,000 megawatts of electricity, hydropower provides about one-fifth of the world's electricity. In Central and South America hydropower provides about 60 per cent of electricity; in Asia this figure is about 15 per cent. Itaipu, on the border of Brazil and Paraguay, is the most powerful hydropower dam built to date, with a capacity of 12,600 megawatts of electricity. It cost $20 billion to build. When finished, China's Three Gorges Dam will produce about 18,200 megawatts of electricity and may cost as much as $75 billion. Thus hydroelectric dams are the most expensive energy projects in the world.

No major river in the world is without existing or planned hydroelectric dams. Nearly four-fifths of the discharge of the largest rivers in Canada, Europe, the former Soviet Union, and the United States are strongly or moderately affected by flow regulation, diversions, and the fragmentation of river channels by dams. More than 500,000 square kilometres—the area of Spain-have been inundated by dam reservoirs world-wide, though not all for hydropower. Globally, about 200 cubic kilometres of water a year-about 7 per cent of the freshwater consumed by human activities-are evaporated from the surface of reservoirs due to their increased exposed surface area. Thus dams affect Earth at scales rivalling other major human activities, such as urbanisation and road building.

Direct Human Impacts

During the 20th century 30-60 million people were flooded off their lands by dams. The World Bank, using Chinese government figures, estimates that 10.2 million people were displaced by reservoirs in China between 1950 and 1989. Given that a number of major dams are under construction or planned in developing countries, there will be no slackening in the pace of population displacement. China's Three Gorges Dam, for example, is expected to displace more than 1 million people, and the proposed Pa Mong Dam between Lao PDR and Thailand is expected to displace more than 500,000.

Large population resettlements can have a number of direct social and health impacts. The social and cultural stress, loss of income, disruption of traditional support services, and other problems facing displaced population often lead to lowered health status. Even when efforts are made to resettle people in new areas, it is difficult to locate land of similar productivity because other groups are likely to already occupy the best areas. Some 13,500 people have been swept to their deaths by the 200 or so dams (outside China) that have collapsed or been over topped in the 20th century. In 1975 in Henan, China, about 230,000 people died from a series of dam bursts.

Disease can spread from vectors that thrive in secondary dam systems, such as irrigation canals and even dam reservoirs. Mosquitoes carrying malaria, for example, have thrived in conditions created by dams. The parasitic disease schistosomiasis has also become more prevalent through the creation of habitats for snails that act as the disease vector. Nearby population, for example, suffered nearly universal infection after several large African dams were filled, including Aswan (Egypt), Akosombo (Ghana), and Sennar (Sudan).

Ecosystem Impacts

An internal survey of World Bank hydroelectric dam projects found that 58 per cent were planned and built without any consideration of downstream impacts-even when these impacts could be predicted to cause coastal erosion, pollution, and other problems.

Dams and greenhouse gases. The work assessing the impacts of dams on greenhouse gas emissions is incomplete, but some estimates have been made. The most immediate changes are in the carbon flow between the flooded vegetation and the atmosphere. The decomposition of plants and soils causes the gradual release of their stored carbon.

From a greenhouse gas standpoint, it might be thought that vegetation decaying in a reservoir would be no worse than the same amount of deforestation. Because of the low-oxygen conditions near and in the bottoms of many reservoirs, however, relative to deforestation a larger fraction of the biomass carbon is likely to be released as methane rather than as carbon dioxide. Since methane is a much more powerful greenhouse gas than carbon dioxide, the global warming impacts are greater than the same amount of carbon released as carbon dioxide. The peak greenhouse gas emissions, however, are unlikely to rival those of a similarly sized fossil power plant, emissions from which would not decrease with age like those from a reservoir.

Nuclear Fuels

There are two main environmental concerns about nuclear power, both mostly with regard to its potential impacts on human health. One involves the highly radioactive products produced by nuclear fission inside power reactors. Such products require careful management at the reactor and during and after disposal. The other concern revolves around the weapons-grade plutonium or uranium that might be clandestinely derived from the nuclear power fuel cycle to make bombs or other weapons of mass destruction by nations or subnational groups.

The routine (non-accidental) emissions of pollutants from the harvesting, processing, and conversion of nuclear fuels are not negligible. And more than many technologies, they are vulnerable to being enhanced by mismanagement. Still, the impacts of these emissions are generally substantially less than those involved with producing power with current coal technologies, the chief competitor in many areas.

Although involving different pollutants, routine emissions from nuclear power systems are probably no more dangerous than those from new natural gas power systems-with the important exception of carbon dioxide, which is not produced by nuclear power. If public concerns about reactor safety, proliferation, and waste disposal can be satisfied, nuclear power may be able to play a significant role in decarbonising the world energy system in the next 50 years.

REGIONAL SCALE PROBLEMS

Nested between local-scale issues-such as the health effects of urban pollution-and global-scale issues-such as climate change-are a number of regional-scale problems that affect human health and ecosystems over areas the size of countries and continents. The most important regional-scale issues are acid deposition, tropospheric ozone, and suspended fine particles. Matched with the regional spatial scale is a temporal scale that requires air pollutants to remain aloft for periods ranging from days to weeks and thereby be transported by prevailing winds and transformed by chemical reactions. Gases and fine particles meet this criterion; larger particles (greater than 1 micron or so in diameter) tend to settle out quickly and are considered contributors to local, rather than regional, impacts. Fine particles may be solid (such as elemental 'black' carbon) or liquid (such as aerosol droplets).

Contributing to regional pollution are a number of precursor species, most of which are generated by the use of fossil fuels and biofuels. Prominent among them are sulphur dioxide (SO_2) and nitrogen oxides (NO_x). Sulphur dioxide is released during the combustion of the sulphur found naturally in fossil fuels, while nitrogen oxides originate either as fuel nitrogen or as atmospheric nitrogen oxidised during combustion.

Other species of importance are particulate matter (PM), carbon monoxide (CO), methane (CH_4), and non-methane volatile organic compounds (NMVOC), released during incomplete combustion and other activities.

Ammonia (NH_3) is a significant regional pollutant, but fuel combustion is not its primary source.

When emissions of these primary species are released into the atmosphere, they form a complex, reactive 'soup,' the chemical and physical behaviour of which is determined by such factors as temperature, humidity, and sunlight. The primary species are transported and deposited, influencing the health of humans and of natural ecosystems. But these primary species are also transformed into secondary species-such as sulphate, nitrate, acids, ozone, and miscellaneous organic compounds-that can have effects even more damaging than their precursors and in areas far removed from the primary sources. This can lead to transboundary problems, where a country or region has little control over the emissions that damage its environment.

Emissions and Energy

The emissions are partitioned into those derived from energy-related activities (including combustion, extraction, processing, and distribution of fossil fuels and biofuels) and those derived from non-energy activities. Non-anthropogenic sources (volcanoes, soils) are not included.

Energy activities account for 82 per cent of anthropogenic emissions of sulphur dioxide and 76 per cent for nitrogen oxides. Energy activities play a less dominant role for the three other species-56 per cent for non-methane volatile organic compounds, 46 per cent for carbon monoxide, and 34 per cent for methane. The smaller role of energy in emissions of these three species reflects the important contributions of deforestation, savannah burning, and agricultural waste burning in the generation of products of incomplete combustion, coupled with rice cultivation and enteric fermentation in the case of methane.

Sulphur dioxide and nitrogen oxides play a role in the formation of acid deposition, because they can be transformed to acids in the atmosphere. The transformation products are fine particles, solid or aerosol, in the form of sulphates and nitrates. Carbon monoxide is an important regional atmospheric pollutant from several perspectives. It acts as an indirect greenhouse gas with a potential for global warming on a 20-year time horizon of about 4.5 due to its influence on the atmospheric lifetime of methane.

Non-methane volatile organic compounds consist of a variety of chemical species. In China, for example, the mix of organic compounds is 46 per cent paraffins, 32 per cent olefins, 21 per cent aromatics, and 1 per cent aldehydes. These compounds are important in the chemistry of the atmosphere because of their influence on the formation and destruction of ozone and methane. Non-methane volatile organic compounds are a product of the incomplete combustion of fossil fuels, biofuels, and other carbonaceous materials. They are also emitted during the extraction, processing, and handling of gaseous and liquid fossil fuels. And they are released through the evaporation of miscellaneous organic products in industry and households.

Ammonia is a significant component of regional emissions. Being an alkaline substance, it neutralises acids in the atmosphere. But once it is deposited on land, it can be converted to acid through biochemical processes in the soil. Ammonia emissions are largely derived from animal waste, fertiliser application, and fuel combustion. In 1990 energy-related activities accounted for just 5 per cent of global ammonia emissions—2.7 of 52.0 teragrams.

Most ammonia emissions are from Asia and other developing countries, due to the rural nature of these countries, the intensive use of fertiliser for food production, and the heavy use of fossil fuels. In 1990 ammonia emissions in Asia were 22.5 teragrams, compared with 3.5 teragrams in Western Europe and 4.1 teragrams in North America.

Future Emissions

Sulphur Dioxide

The latest energy projections indicate that global sulphur dioxide emissions will likely stay roughly constant between 1990 and 2020, at about 59 teragrams of sulphur. This 'middle-course' scenario incorporates modest economic growth, continued reliance on fossil fuels, and the elimination of trade barriers. At the regional level, however, a distinctive pattern emerges for all the important species. Emissions will decline in the industrialised regions of the Northern hemisphere. Europe, the former Soviet Union, North America-and increase sharply the developing regions of the Southern hemisphere and the Far East-Latin America, Africa, Asia. In Western Europe strong national environmental policies, changes in national energy policies, and implementation of the 1985 Helsinki Protocol and 1994 Oslo Protocol (under the 1979 Convention on Long-range Transboundary Air Pollution) have driven down sulphur dioxide emissions. As a result the region could see a 60 per cent drop in sulphur dioxide emissions between 1990 and 2020.

The problem of sulphur dioxide emissions has shifted to the developing world, with emissions in Latin America, Africa, and the Middle East expected to increase by about 30 per cent between 1990 and 2020. The main problem region is Asia, where emissions are already high-with 17 teragrams of sulphur emissions in 1990-and could double by 2020. If that happens, Asia will account for 58 per cent of global emissions, much of them from China.

Driven by a similar energy forecast, the model projects that Asia's sulphur dioxide emissions in 2020 will be bounded by an upper value of 40 teragrams of sulphur (under the assumption of no further control policies beyond those in place in 1990- the NFC scenario) and a lower value of 6 teragrams of sulphur (with the assumption of very tight controls, similar to those in Western Europe). A mid-range estimate is 20 teragrams of sulphur, to be achieved if all large, new facilities are fitted with flue-gas desulphurisation units and other fossil-fuel users switch to low-sulphur fuels.

China continues to be the largest contributor to Asia's sulphur dioxide emissions, emitting about half of the continental total. But the establishment in 1997 of China's Two Control Zone policy for sulphur dioxide emissions has generated optimism that emissions will not grow as fast as once thought. Emissions of 11.9 teragrams of sulphur in 1995 are used as a baseline, and the plan is to limit national emissions to 12.3 teragrams of sulphur in 2000 by capping emissions in certain provinces at their 1995 levels. While implementation and enforcement questions linger, there is a commitment at the highest level in China to holding down sulphur dioxide emissions.

The official (but undocumented) estimate for China's emissions in 2020 is 19.5 teragrams of sulphur. With rapid growth in Asia, many of the coal-fired plants projected to be needed after 2000 have yet to be built. Thus the opportunity exists to fit these plants with emission controls or lower-emission technology at the time of construction. The incremental cost of emission reduction is then the only hurdle to be overcome-though it is a high hurdle ($25 billion a year for the ACT scenario, rising to $65 billion a year for the BAT scenario).

Substitution of natural gas for coal is an attractive interim measure, and any possibilities for increasing the efficiency of energy use and moving towards renewable energy would reduce emissions of sulphur, nitrogen, and other species. The ecologically driven scenario, for example, would lower global 2020 emissions from 59 teragrams of sulphur to 34 teragrams.

Nitrogen Oxides

The situation for nitrogen oxides emissions is even more challenging, because of the added emissions from transportation. Though nitrogen emissions were not estimated by Naki´cenovi´c, Grübler, and McDonald, other analyses suggest a regional pattern similar to that of sulphur dioxide. An earlier study, Energy for Tomorrow's World, which was more optimistic about economic growth, forecast a 13 per cent increase in global emissions of nitrogen oxide between 1990 and 2020 (from 24 teragrams of nitrogen to 27 teragrams) under case B assumptions.

The increase in Asia was 70 per cent (from 6.8 teragrams of nitrogen to 11.5 teragrams). Use of the RAINS-ASIA model, with its daunting view of the growth of fossil-fuel-based energy systems in Asia, yields an estimated increase of more than 300 per cent in this period.

Carbon monoxide and non-methane volatile organic compounds. Though there are no published projections for emissions of carbon monoxide and non-methane volatile organic compounds, carbon monoxide emissions are unlikely to increase in Asia, because inefficient combustion of biofuels will fall and inefficient vehicles will be replaced.

Acid Deposition Process

Acid deposition-or acid precipitation in its 'wet' form-is perhaps the most important regional-scale manifestation of energy generation through fuel combustion. Acid deposition occurs when sulphur dioxide and nitrogen oxides are oxidised in the atmosphere to-sulphuric acid (H_2SO_4) and nitric acid (HNO_3), respectively, and dissolved in rainwater. Clean rainwater is naturally acidic, with a pH of about 5.6. In the industrialised regions of Europe, North America, and Asia, rainfall pH values of 4.0-6.0 are common-and values as low as 3.0 have been measured in isolated events.

Acid deposition is a problem because it causes damage to natural and human-made surfaces with which it comes into contact. If soils contain insufficient alkali to neutralise the acid, damage can be caused to vegetation, particularly sensitive tree species and agricultural crops. Lakes can become acidified, leading to the demise of fish population. Over time the entire natural structure and function of ecosystems can change. Manufactured materials can be attacked: metal surfaces rust, and alkaline materials like concrete, limestone, and marble are eroded.

In Europe forest damage has long been attributed to acid deposition. Despite emission reductions, the health of European forests still seems to be deteriorating. In a 1995 survey of 117,000 trees in 30 countries, more than 25 per cent showed signs of significant defoliation, and more than 10 per cent showed significant leaf discoloration. Both direct and indirect effects of air pollution, of which acid deposition is but one part, are considered the cause.

Surveys of forest soils show that, while sulphur deposition has dropped drastically since the 1970s, nitrogen deposition is still high, impairing soil chemistry and nutrient status. For acidification of surface waters, there appears to be an overall improvement, probably as a result of reductions in acid deposition. With projected reductions in sulphur and nitrogen emissions through

2020, continued progress is expected towards healthier ecosystems in Europe.

North America has seen significant reductions in the sulphate concentration and pH of precipitation as a result of the 1990 amendments to the Clean Air Act. Reductions in nitrate concentration have not been observed, however, because requirements for lower nitrogen oxide emissions did not go into effect until 1996. On the whole, it is too early to tell if there has been significant improvement in the health of ecosystems. There is evidence of recovery in some New England lakes, but the U.S. Environmental Protection Agency has reported that additional reductions in sulphur and nitrogen deposition will be needed to fully restore the health of sensitive Adirondack lakes.

High-elevation spruce fir forests in the eastern United States continue to show signs of damage. But, as in Europe, there is reason to hope for improvement. Asia is the region of greatest concern. Acid deposition is being reported throughout Asia, with many areas receiving levels that exceed the carrying capacity of their soils. Longrange transport is receiving scientific and political attention as countries receive increasing pollution from neighbouring and even distant countries. By far the worst episodes of acid deposition occur in southwestern China. Average rainwater pH values of 4.0-5.0 are observed in the Sichuan Basin, and values below 3.0 have been recorded in individual episodes. Atmospheric conditions in Sichuan and Guizhou provinces, with weak winds and frequent temperature inversions, are conducive to high pollutant concentrations. Emissions are also high there because of the widespread burning of high-sulphur coal in small stoves and medium-size industrial boilers.

Southwestern China has seen damage from acid deposition. Sulphur deposition levels are more than 10 grams of sulphur per square metre per year, making the situation comparable to the worst parts of the former

Czechoslovakia in the 1960s and 1970s. Zhao and Xiong report the following effects in the vicinity of Chongqing and the provinces of Sichuan and Guizhou:

1. A 50 per cent dieback of pine forests on Nanshan Mountain, about 20 kilometres from Chongqing, attributed to acid deposition and air pollution.
2. A more than 50 per cent reduction in biomass production in commercial spruce forests in areas experiencing rain with a pH of less than 4.5.
3. A yellowing of rice in large areas near Chongqing after rainfalls with a pH of less than 4.5.

Seip and others sampled soil water and stream water in a 7-hectare catchment near Guiyang in Guizhou Province, about 350 kilometres south of Chongqing. Sulphate concentrations were very high, pH values were as low as 4.3, and aluminium concentrations were elevated. Despite these factors, no apparent damage to vegetation was observed. It appears that neutralisation of acid inputs by deep soils and underlying bedrock may be averting ecosystem damage. Because of the heterogeneity of Chinese soils, however, local acidification and damage may be occurring in sensitive areas that have not been studied. A more recent survey of acidification in China by the Norwegian Institute for Water Research reported severe effects of acid deposition on soils, water bodies with high loadings showing typical signs of acidification, and observed effects on surface water organisms.

Zhao and Xiong describe some of the severe materials damage observed in Chongqing: "Metal structures are scraped of rust and painted once every 1-3 years. Shells of buses are generally replaced every 1-2 years. Structural components made of stainless steel become rusty after a few years. Some concrete works built in the 1950s have corroded in such a manner that the gravel is exposed. It is estimated that the corrosion depth reaches 0.5 cm in less than 30 years".

In northern China, by contrast, rainwater pH values are typically 6.0-7.0. Although emissions are high in many parts of northern China, meteorological conditions are more conducive to pollutant dispersion, and windblown dust from central Asian deserts tends to neutralise the acidity. The line delineating acid precipitation in China (pH of 5.6) extends just west of Beijing and along the eastern edge of the Greater Khingan mountain range. Since 1982 the area receiving acid deposition may have expanded by 600,000-700,000 square kilometres. Acidification is responsible for much of the air pollution-related damage in China, though the relative roles of acid rain, dry deposition of sulphur dioxide, nitrates, particulates, ozone, and other factors have not been determined. Areas with lower rain acidity see much less damage than Chongqing and neighbouring cities of southwestern China. Acid rain damage to crops, forests, materials, and human health in China in 1995 is estimated to total more than $13 billion.

In Asia there is also considerable concern about the fate of cultural materials as pollution levels rise. Concerns about the deterioration of the Taj Mahal were raised as far back as 1981. Throughout Asia, cultural buildings and monuments made of alkaline-based materials are vulnerable to attack. Glass, paper, textiles, and archives are also subject to accelerated deterioration in the warm, moist, polluted atmospheres of Asia. These problems are greatly under-appreciated and should be given high priority in future research before rich areas of cultural heritage are destroyed. Finally, although not yet major emitters, Sub-Saharan Africa and Latin American have the potential for significant sulphur emissions as fossil fuel use increases.

Tropospheric Ozone

Ozone is an important air pollutant that can cause

damage to crops, trees, and human health. It is a major component of the harmful smog that forms in urban areas during periods of high temperature, intense solar radiation, low wind speed, and an absence of precipitation. In the polluted air mass, ozone is produced by a complex set of chemical reactions involving nitrogen oxides and nonmethane volatile organic compounds. North America and Europe are developing coordinated strategies to reduce emissions of ozone precursors and thereby reduce some of the health and ecosystem damage attributed to it. Although there is still progress to be made in these regions, it is again in Asia that concern is greatest.

Episodes of high ozone concentrations are now common in the megacities (cities containing more than 10 million people) of southern Asia that have industrial emissions (producing volatile organic compounds), transportation (producing nitrogen oxides), and conducive climates-Bangkok (Thailand), Hong Kong (China), Mumbai (India), and Shanghai (China), to name a few. In addition, the formation and transport of regional ozone have been observed in measurement campaigns such as PEM-West.

Ozone concentrations were observed to be regionally enhanced by photochemical activity in continental air masses passing through areas with high nitrogen oxides emissions. The potential effects of elevated ozone concentrations on human health and crop production in Asia are just beginning to be explored. Studies in the West have established that crop yields are depressed by repeated exposures to ozone levels above 50-70 parts per billion; these concentrations are exceeded in fall and winter throughout large areas of southern China.

There is concern that damage to winter wheat and other crops in the Yangtze Delta may endanger China's ability to meet increasing food demands. These analyses are still in their infancy, however, and much more work is needed on meteorological analysis, the gathering of monitoring data, field studies on crop responses to elevated concentrations, and regional assessments of economic impact. Until more of this work is done in Asia, a definitive statement cannot be made about the relationship between regional emissions of nonmethane volatile organic compounds and nitrogen oxides and impacts on human health and vegetation.

Climate Change

In the early 1990s it was recognised that sulphate aerosols can influence the global climate by scattering and absorbing incoming solar radiation and hence exerting a cooling effect. The role of sulphate aerosols has now been clarified. Indeed, sulphate aerosols contribute negative radiative forcing (of about -0.4 watts per square metre) that offsets the positive forcing of carbon dioxide and other greenhouse gases. Hence a reduction in sulphur dioxide or nitrogen oxide emissions would be expected to reduce sulphate aerosol concentrations and increase the potential for global warming. The radiative forcing is spatially in homogeneous, with values as large as -11 watts per square metre over heavily polluted areas such as Central Europe and eastern China.

Lal and others have suggested that sulphate aerosols can also interfere with local climates. The cooling effect of the aerosol haze reduces the difference between land and sea temperatures and weakens the monsoon. In addition, the cooler land surface reduces evaporation and lowers the amount of water vapour in the atmosphere. The authors estimate that sulphate aerosols in the Asian atmosphere will reduce monsoon rainfall over India and parts of China by the middle of this century.

9

BIODIVERSITY AND CONSERVATION

Biodiversity is the variation of taxonomic life forms within a given ecosystem, biome or for the entire Earth. Biodiversity is often a measure of the health of biological systems.

The most straightforward definition is "variation of life at all levels of biological organisation". A second definition holds that biodiversity is a measure of the relative diversity among organisms present in different ecosystems. "Diversity" in this definition includes diversity within a species and among species, and comparative diversity among ecosystems.

A third definition that is often used by ecologists is the "totality of genes, species, and ecosystems of a region". An advantage of this definition is that it seems to describe most circumstances and present a unified view of the traditional three levels at which biodiversity has been identified:

1. Genetic diversity—diversity of genes within a species. There is a genetic variability among the population and the individuals of the same species.
2. Species diversity—diversity among species in an ecosystem.

3. Ecosystem diversity—diversity at a higher level of organisation, the ecosystem. To do with the variety of ecosystems on Earth.

The 1992 United Nations Earth Summit in Rio de Janeiro defined "biodiversity" as "the variability among living organisms from all sources, including, 'inter alia', terrestrial, marine, and other aquatic ecosystems, and the ecological complexes of which they are part: this includes diversity within species, between species and of ecosystems". This is, in fact, the closest thing to a single legally accepted definition of biodiversity, since it is the definition adopted by the United Nations Convention on Biological Diversity.

If the gene is the fundamental unit of natural selection, according to E. O. Wilson, the real biodiversity is genetic diversity. For geneticists, *biodiversity* is the diversity of genes and organisms. They study processes such as mutations, gene exchanges, and genome dynamics that occur at the DNA level and generate evolution. For ecologists, biodiversity is also the diversity of durable interactions among species. It not only applies to species, but also to their immediate environment (biotope) and their larger ecoregion. In each ecosystem, living organisms are part of a whole, interacting with not only other organisms, but also with the air, water, and soil that surround them.

MEASUREMENT OF BIODIVERSITY

Biodiversity is a broad concept, so a variety of objective measures have been created in order to empirically measure biodiversity. Each measure of biodiversity relates to a particular use of the data.

For practical conservationists, this measure should quantify a value that is broadly shared among locally affected people. For others, a more economically defensible definition should allow the ensuring of continued possibilities for both adaptation and future use by people, assuring environmental sustainability.

As a consequence, biologists argue that this measure is likely to be associated with the variety of genes. Since it cannot always be said which genes are more likely to prove beneficial, the best choice for conservation is to assure the persistence of as many genes as possible. For ecologists, this latter approach is sometimes considered too restrictive, as it prohibits ecological succession.

Biodiversity is usually plotted as taxonomic richness of a geographic area, with some reference to a temporal scale. Whittaker described three common metrics used to measure species-level biodiversity, encompassing attention to species richness or species evenness:

1. Species richness - the least sophisticated of the indices available.
2. Simpson index.
3. Shannon index.

There are three other indices which are used by ecologists:

1. Alpha diversity refers to diversity within a particular area, community or ecosystem, and is measured by counting the number of taxa within the ecosystem (usually species)
2. Beta diversity is species diversity between ecosystems; this involves comparing the number of taxa that are unique to each of the ecosystems.
3. Gamma diversity is a measure of the overall diversity for different ecosystems within a region.

BIODIVERSITY DISTRIBUTION

Biodiversity is not distributed evenly on Earth. It is

consistently richer in the tropics and in other localised regions such as the California Floristic Province. As one approaches polar regions one generally finds fewer species. Flora and fauna diversity depends on climate, altitude, soils and the presence of other species. In the year 2006 large numbers of the Earth's species are formally classified as rare or endangered or threatened species; moreover, most scientists estimate that there are millions more species actually endangered which have not yet been formally recognised. About 40 per cent of the 40,177 species assessed using the IUCN Red List criteria, are now listed as threatened species with extinction—a total of 16,119 species.

A biodiversity hotspot is a region with a high level of endemic species. These biodiversity hotspots were first identified by Dr. Norman Myers in two articles in the scientific journal *The Environmentalist*. Dense human habitation tends to occur near hotspots. Most hotspots are located in the tropics and most of them are forests.

Brazil's Atlantic Forest is considered a hotspot of biodiversity and contains roughly 20,000 plant species, 1350 vertebrates, and millions of insects, about half of which occur nowhere else in the world. The island of Madagascar including the unique Madagascar dry deciduous forests and lowland rainforests possess a very high ratio of species endemism and biodiversity, since the island separated from mainland Africa 65 million years ago, most of the species and ecosystems have evolved independently producing unique species different than other parts of Africa. Many regions of high biodiversity (as well as high endemism) arise from very specialised habitats which require unusual adaptation mechanisms.

BIODIVERSITY AND EVOLUTION

Biodiversity found on Earth today is the result of 4 billion years of evolution. The origin of life has not been definitely established by science, though evidence suggests that life may already have been well-established a few 100 million years after the formation of the Earth. Until approximately 600 million years ago, all life consisted of bacteria and similar single-celled organisms.

The history of biodiversity during the Phanerozoic (the last 540 million years), starts with rapid growth during the Cambrian explosion—a period during which nearly every phylum of multicellular organisms first appeared. Over the next 400 million years or so, global diversity showed little overall trend, but was marked by periodic, massive losses of diversity classified as mass extinction events.

The apparent biodiversity shown in the fossil record suggests that the last few million years include the period of greatest biodiversity in the Earth's history. However, not all scientists support this view, since there is considerable uncertainty as to how strongly the fossil record is biased by the greater availability and preservation of recent geologic sections. Some argue that corrected for sampling artifacts, modern biodiversity is not much different from biodiversity 300 million years ago. Estimates of the present global macroscopic species diversity vary from 2 million to 100 million species, with a best estimate of somewhere near 13-14 million, the vast majority of them arthropods.

Most biologists agree however that the period since the emergence of humans is part of a new mass extinction, the Holocene extinction event, caused primarily by the impact humans are having on the environment. At present, the number of species estimated to have gone extinct as a result of human action is still far smaller than are observed during the major mass extinctions of the geological past. However, it has been argued that the present rate of extinction is sufficient to create a major mass extinction in less than 100 years. Others dispute this and suggest that the present rate of extinctions could be sustained for many thousands of years before the loss of biodiversity matches the more than 20% losses seen in past global extinction events.

New species are regularly discovered (on average between 5-10,000 new species each year, most of them insects) and many, though discovered, are not yet classified (estimates are that nearly 90 present of all arthropods are not yet classified). Most of the terrestrial diversity is found in tropical forests.

BENEFITS OF BIODIVERSITY

There are a multitude of benefits of biodiversity in the sense of one diverse group aiding another such as:

Resistance to Catastrophe

Monoculture, the lack of biodiversity, was a contributing factor to several agricultural disasters in history, including the Irish Potato Famine, the European wine industry collapse in the late 1800s, and the US Southern Corn Leaf Blight epidemic of 1970. Higher biodiversity also controls the spread of certain diseases as viruses will need adapt to infect different species.

Food and Drink

Biodiversity provides food for humans. Although about 80 per cent of the food supply comes from just 20 kinds of plants, humans use at least 40,000 species of plants and animals a day. Many people around the world depend on these species for their food, shelter, and clothing. There is untapped potential for increasing the range of food products suitable for human consumption, provided that the high present extinction rate can be stopped.

Medicines

A significant proportion of drugs are derived, directly or indirectly, from biological sources; in most cases these medicines can not presently be synthesised in a laboratory setting. About 40 present of the pharmaceuticals used in the US are manufactured using natural compounds found in plants, animals, and microorganisms. Moreover, only a small proportion of the total diversity of plants has been thoroughly investigated for potential sources of new drugs. Many drugs are also derived from microorganisms.

Ranges of Industrial Materials

A wide range of industrial materials are derived directly from biological resources. These include building materials, fibers, dyes, resins, gums, adhesives, rubber and oil. There is enormous potential for further research into sustainably utilising materials from a wider diversity of organisms.

Intellectual Value

Through the field of bionics, a lot of technological advancement has been done which may not have been the case without a rich biodiversity.

Better Crop-varieties

For some foodcrops and other economic crops, wild varieties of the domesticated species can be reintroduced to form a better variety than the previous (domesticated) species. The economic impact is gigantic, for even crops as common as the potato (which was bred through only one variety, brought back from the Inca), a lot more can come from these species. Wild varieties of the potato will all suffer enormously through the effects of climate change.

A report by the Consultative Group on International Agricultural Research (CGIAR) describes the huge economic loss. Rice, which has been improved for thousands of years by humans, can through the same process regain some of its nutritional value that has been lost since (a project is already being carried out to do just this).

Crop diversity is also necessary to help the system recover when the dominant crop type is attacked by a disease:

1. The Irish potato blight of 1846, which killed a million people and forced another million to emigrate, was the result of planting only two potato varieties, both of which were vulnerable.
2. When the rice grassy stunt virus struck rice fields from Indonesia to India in the 1970s, 6273 varieties were tested. Only one was luckily found to be resistant, a relatively feeble Indian variety, known to science only since 1966, with the desired trait. It was hybridised with other varieties and now widely grown.
3. In 1970, coffee rust attacked coffee plantations in Sri Lanka, Brazil, and Central America. A resistant variety was found in Ethiopia, coffee's presumed homeland, which saved the day.

HINDRANCES

There are also many impracticalities in saving all the species currently on earth:

Public Funds

Humans have always expanded and developed their

territory throughout history. An active approach is the only way to hault the expansion but this often requires public funds. Currently the United States Environmental Protection Agency has an annual budget of $7.3 billion.

Natural Resources

Many times natural resources cannot be exploited due to environmental protection acts. Fine woods from South America and oil from Alaska are prime examples, where even minor human activities are deemed too disruptive to the overall health and biodiversity of an area to be allowed.

Preservation of Invertebrate

Biodiverstiy is most closely known to the public as a loss of animals with a backbone when in fact there exist 20 times that number of insects 5 times as many flowering plants. While a few of these species may be highly valuable to the human race for the above reasons, the vast majority are often completely unknown to anyone but specialists. In fact it is often estimated that less than half and perhaps less than two-thirds of earth organisms have even been identified.

BIODIVERSITY AND ITS THREATS

During the last century, erosion of biodiversity has been increasingly observed. Some studies show that about one eighth known plant species is threatened with extinction. Some estimates put the loss at up to 140,000 species per year (based on Species-area theory) and subject to discussion. This figure indicates unsustainable ecological practices, because only a small number of species come into being each year. Almost all scientists acknowledge that the rate of species loss is greater now than at any time in human history, with extinctions occurring at rates hundreds of times higher than background extinction rates.

Habitat Destruction

Most of the species extinctions from 1000 AD to 2000 AD are due to human activities, in particular destruction of plant and animal habitats. Raised rates of extinction are being driven by human consumption of organic resources, especially related to tropical forest destruction. While most of the species that are becoming extinct are not food species, their biomass is converted into human food when their habitat is transformed into pasture, cropland, and orchards.

It is estimated that more than 40 present of the Earth's biomass is tied up in only the few species that represent humans, livestock and crops. Because an ecosystem decreases in stability as its species are made extinct, these studies warn that the global ecosystem is destined for collapse if it is further reduced in complexity. Factors contributing to loss of biodiversity are: overpopulation, deforestation, pollution (air pollution, water pollution, soil contamination) and global warming or climate change, driven by human activity. These factors, while all stemming from overpopulation, produce a cumulative impact upon biodiversity.

Some characterise loss of biodiversity not as ecosystem degradation but by conversion to trivial standardised ecosystems (e.g., monoculture following deforestation). In some countries lack of property rights or access regulation to biotic resources necessarily leads to biodiversity loss (degradation costs having to be supported by the community).

A September 14, 2007 study conducted by the National Science Foundation found that biodiversity and genetic diversity are dependent upon each other—that diversity within a species is necessary to maintain diversity among species, and vice versa. According to the lead researcher in the study, Dr. Richard Lankauof, "If any one type is removed from the system, the cycle can break down, and the community becomes dominated by a single species."

Exotic or Invasive Species

The rich diversity of unique species across many parts of the world exist only because they are separated by barriers, particularly large rivers, seas, oceans, mountains and deserts from other species of other land masses, particularly the highly fecund, ultra-competitive, generalist "super-species". These are barriers that could never be crossed by natural processes, except for many millions of years in the future through continental drift. However humans have invented ships and airplanes, and now have the power to bring into contact species that never have met in their evolutionary history, and on a time scale of days, unlike the centuries that historically have accompanied major animal migrations.

The widespread introduction of exotic species by humans is a potent threat to biodiversity. When exotic species are introduced to ecosystems and establish self-sustaining population, the endemic species in that ecosystem, that have not evolved to cope with the exotic species, may not survive. The exotic organisms may be either predators, parasites, or simply aggressive species that deprive indigenous species of nutrients, water and light. These exotic or invasive species often have features, due to their evolutionary background and new environment, that make them highly competitive; able to become well-established and spread quickly, reducing the effective habitat of endemic species.

Process of Genetic Pollution

Purebred naturally evolved region specific wild species can be threatened with extinction in a big way through the process of Genetic Pollution i.e. uncontrolled hybridisation, introgression and Genetic swamping which leads to homogenisation or replacement of local genotypes as a result of either a numerical and/or fitness advantage of introduced plant or animal. Nonnative species can bring about a form of extinction of native plants and animals by hybridisation and introgression either through purposeful introduction by humans or through habitat modification, bringing previously isolated species into contact. These phenomena can be especially detrimental for rare species coming into contact with more abundant ones where the abundant ones can interbreed with them swamping the entire rarer gene pool creating hybrids thus driving the entire original purebred native stock to complete extinction.

Attention has to be focused on the extent of this under appreciated problem that is not always apparent from morphological (outward appearance) observations alone. Some degree of gene flow may be a normal, evolutionarily constructive process, and all constellations of genes and genotypes cannot be preserved however, hybridisation with or without introgression may, nevertheless, threaten a rare species' existence.

Use of Conventional Hybridisation

In agriculture and animal husbandry, green revolution popularised the use of conventional hybridisation to increase yield many folds by creating "high-yielding varieties". Often the handful of breeds of plants and animals hybridised originated in developed countries and were further hybridised with local varieties, in the rest of the developing world, to create high yield strains resistant to local climate and diseases.

Local governments and industry since have been pushing hybridisation with such zeal that several of the wild and indigenous breeds evolved locally over thousands of years having high resistance to local extremes in climate and immunity to diseases etc. have already become extinct or are in grave danger of becoming so in the near future. Due to complete disuse because of un-profitability and uncontrolled intentional, compounded with unintentional cross-pollination and crossbreeding (genetic pollution) formerly huge gene pools of various wild and indigenous breeds have collapsed causing widespread genetic erosion and genetic pollution resulting in great loss in genetic diversity and biodiversity as a whole.

A genetically modified organism (GMO) is an organism whose genetic material has been altered using the genetic engineering techniques generally known as recombinant DNA technology. Genetically Modified (GM) crops today have become a common source for genetic pollution, not only of wild varieties but also of other domesticated varieties derived from relatively natural hybridisation.

It is being said that genetic erosion coupled with genetic pollution is destroying that needed unique genetic base thereby creating an unforeseen hidden crisis which will result in a severe threat to the food security for the future when diverse genetic material will cease to exist to be able to further improve or hybridise weakening food crops and livestock against more resistant diseases and climatic changes.

CONSERVATION OF BIOLOGICAL DIVERSITY

The conservation of biological diversity has become a global concern. Although not everybody agrees on extent and significance of current extinction, most consider biodiversity essential. There are basically two main types of conservation options, in-situ conservation and ex-situ conservation. In-situ is usually seen as the ideal conservation strategy. However, its implementation is sometimes infeasible. For example, destruction of rare or endangered species' habitats sometimes requires ex-situ conservation efforts.

Furthermore, ex-situ conservation can provide a backup solution to in-situ conservation projects. Some believe both types of conservation are required to ensure proper preservation. An example of an in-situ conservation effort is the setting-up of protection areas. Examples of ex-situ conservation efforts, by contrast, would be planting germplasts in seedbanks, or growing the Wollemi Pine in nurseries. Such efforts allow the preservation of large population of plants with minimal genetic erosion.

At national levels a Biodiversity Action Plan is sometimes prepared to state the protocols necessary to protect an individual species. Usually this plan also details extant data on the species and its habitat. In the USA such a plan is called a Recovery Plan. The threat to biological diversity was among the hot topics discussed at the UN World Summit for Sustainable Development, in hope of seeing the foundation of a Global Conservation Trust to help maintain plant collections.

POLITICAL AND JUDICIAL DECISIONS

Biodiversity is beginning to be evaluated and its evolution analysed (through observations, inventories, conservation...) as well as being taken into account in political and judicial decisions.

1. The relationship between law and ecosystems is very ancient and has consequences for biodiversity. It is related to property rights, both private and

public. It can define protection for threatened ecosystems, but also some rights and duties.

2. Law regarding species is a more recent issue. It defines species that must be protected because they may be threatened by extinction. Some people question application of these laws. The U.S. Endangered Species Act is an example of an attempt to address the "law and species" issue.
3. Laws regarding gene pools are only about a century old. While the genetic approach is not new (domestication, plant traditional selection methods), progress made in the genetic field in the past 20 years have led to a tightening of laws in this field. With the new technologies of genetic analysis and genetic engineering, people are going through gene patenting, processes patenting, and a totally new concept of genetic resources. A very hot debate today seeks to define whether the resource is the gene, the organism itself, or its DNA.

The 1972 UNESCO convention established that biological resources, such as plants, were the common heritage of mankind. These rules probably inspired the creation of great public banks of genetic resources, located outside the source-countries.

New global agreements (e.g. Convention on Biological Diversity), now give sovereign national rights over biological resources (not property). The idea of static conservation of biodiversity is disappearing and being replaced by the idea of dynamic conservation, through the notion of resource and innovation.

The new agreements commit countries to conserve biodiversity, develop resources for sustainability and share the benefits resulting from their use. Under new rules, it is expected that bioprospecting or collection of natural products has to be allowed by the biodiversity-rich country, in exchange for a share of the benefits.

Sovereignty principles can rely upon what is better known as Access and Benefit Sharing Agreements (ABAs). The Convention on Biodiversity spirit implies a prior informed consent between the source country and the collector, to establish which resource will be used and for what, and to settle on a fair agreement on benefit sharing. Bioprospecting can become a type of biopiracy when those principles are not respected.

Uniform approval for use of biodiversity as a legal standard has not been achieved, however. At least one legal commentator has argued that biodiversity should not be used as a legal standard, arguing that the multiple layers of scientific uncertainty inherent in the concept of biodiversity will cause administrative waste and increase litigation without promoting preservation goals

BIODIVERSITY AND CONSERVATION

An array of interactions among the various components of biodiversity makes the planet habitable for all species, including humans. Our personal health, and the health of the economy and human society, depends on the continuous supply of various ecological services that would be extremely costly or impossible to replace. These natural services are so varied as to be almost infinite.

Biodiversity Under Threat

When most people think of the dangers besetting the natural world, they think of the threat to other creatures. Declines in the numbers of such charismatic animals as pandas, tigers, elephants, whales, and various species of birds, have drawn world attention to the problem of species at risk. Species have been disappearing at 50-100 times the natural rate, and this is predicted to rise dramatically. Based on current trends, an estimated 34,000 plant and 5,200 animal species, including one in eight of the world's bird species are facing extinction.

For thousands of years we have been developing a vast array of domesticated plants and animals important for food. But this treasure house is shrinking as modern commercial agriculture focuses on relatively few crop varieties, from which we constantly draw. It destabilises ecosystems, and weakens their ability to deal with natural disasters such as floods, droughts, and hurricanes, and with human- caused stresses, such as pollution and climate change. Already, we are spending huge sums in response to flood and storm damage exacerbated by deforestation; such damage is expected to increase due to global warming.

The reduction in biodiversity also hurts us in other ways. Our cultural identity is deeply rooted in the biological environment. Plants and animals are symbols of the world, preserved in flags, sculptures, and other images that define the societies and us. While loss of species has always occurred as a natural phenomenon, the pace of extinction has accelerated dramatically as a result of human activity.

Ecosystems are being fragmented or eliminated, and innumerable species are in decline or already extinct. We are creating the greatest extinction crisis since the natural disaster that wiped out the dinosaurs 65 million years ago. These extinctions are irreversible and, given the dependence on food crops, medicines and other biological resources, pose a threat to our own well- being. It is reckless if not downright dangerous to keep chipping away at our life support system.

It is unethical to drive other forms of life to extinction, and thereby deprive present and future generations of options for their survival and development. Can we save the world's ecosystems, and with them the species we value and the other millions of species, some of which may produce the foods and medicines of tomorrow? The answer will lie in the ability to bring the demands into line with nature's ability to produce what we need and to safely absorb what we throw away. While concern for the environment is constant in history, heightened concern about environmental destruction and loss of species and ecosystems in the seventies led to concerted action.

In 1972, the United Nations Conference on the Human Environment (Stockholm) resolved to establish the United Nations Environment Programme (UNEP). Governments signed a number of regional and international agreements to tackle specific issues, such as protecting wetlands and regulating the international trade in endangered species. These agreements, along with controls on toxic chemicals and pollution, have helped to slow the tide of destruction but have not reversed it. For example, an international ban and restrictions on the taking and selling of certain animals and plants have helped to reduce over- harvesting and poaching.

National Action Plan for Biological Diversity

The Convention on Biological Diversity, as an international treaty, identifies a common problem, sets overall goals and policies and general obligations, and organises technical and financial cooperation. However, the responsibility for achieving its goals rests largely with the countries themselves. Private companies, landowners, fishermen, and farmers take most of the actions that affect biodiversity. Governments need to provide the critical role of leadership, particularly by setting rules that guide the use of natural resources, and by protecting biodiversity where they have direct control over the land and water.

Under the Convention, governments undertake to conserve and sustainable use biodiversity. They are required to develop national biodiversity strategies and action plans, and to integrate these into broader national plans for environment and development. This is particularly important for such sectors as forestry, agriculture, fisheries, energy, transportation and urban planning. Other treaty commitments include:

— Identifying and monitoring the important components of biological diversity that need to be conserved and used sustainable.

— Establishing protected areas to conserve biological diversity while promoting environmentally sound development around these areas.

— Rehabilitating and restoring degraded ecosystems and promoting the recovery of threatened species in collaboration with local residents.

— Respecting, preserving and maintaining traditional knowledge of the sustainable use of biological diversity with the involvement of indigenous peoples and local communities.

— Preventing the introduction of, controlling, and eradicating alien species that could threaten ecosystems, habitats or species.

— Controlling the risks posed by organisms modified by biotechnology.

— Promoting public participation, particularly when it comes to assessing the environmental impacts of development projects that threaten biological diversity.

— Educating people and raising awareness about the importance of biological diversity and the need to conserve it.

— Reporting on how each country is meeting its biodiversity goals.

Issues dealt with under the Convention include:

— Measures and incentives for the conservation and sustainable use of biological diversity.

— Regulated access to genetic resources.
— Access to and transfer of technology, including biotechnology.
— Technical and scientific cooperation.
— Impact assessment.
— Education and public awareness.
— Provision of financial resources.
— National reporting on efforts to implement treaty commitments.

In International Law

It recognises, for the first time, that the conservation of biological diversity is "a common concern of humankind" and is an integral part of the development process. The agreement covers all ecosystems, species, and genetic resources. It links traditional conservation efforts to the economic goal of using biological resources sustainable. It sets principles for the fair and equitable sharing of the benefits arising from the use of genetic resources, notably those destined for commercial use. It also covers the rapidly expanding field of biotechnology, addressing technology development and transfer, benefit-sharing and biosafety. Importantly, the Convention is legally binding; countries that join it are obliged to implement its provisions.

The Convention reminds decision-makers that natural resources are not infinite and sets out a new philosophy for the 21st century, that of sustainable use. While past conservation efforts were aimed at protecting particular species and habitats, the Convention recognises that ecosystems, species and genes must be used for the benefit of humans. However, this should be done in a way and at a rate that does not lead to the long- term decline of biological diversity.

The Convention also offers decision- makers guidance based on the precautionary principle that where there is a threat of significant reduction or loss of biological diversity, lack of full scientific certainty should not be used as a reason for postponing measures to avoid or minimise such a threat. The Convention acknowledges that substantial investments are required to conserve biological diversity. It argues, however, that conservation will bring us significant environmental, economic and social benefits in return.

Philosophy of Biological Diversity

In 1992, the largest-ever meeting of world leaders took place at the United Nations Conference on Environment and Development in Rio de Janeiro, Brazil. An historic set of agreements was signed at the "Earth Summit", including two binding agreements, the Convention on Climate Change, which targets industrial and other emissions of greenhouse gases such as carbon dioxide, and the Convention on Biological Diversity, the first global agreement on the conservation and sustainable use of biological diversity.

The biodiversity treaty gained rapid and widespread acceptance. Over 150 governments signed the document at the Rio conference, and since then more than 175 countries have ratified the agreement.

The Convention has three main goals:

— The conservation of biodiversity,

— Sustainable use of the components of biodiversity, and

— Sharing the benefits arising from the commercial and other utilisation of genetic resources in a fair and equitable way.

The Convention is comprehensive in its goals, and deals with an issue so vital to humanity's future, that it stands as a landmark 10 in the rice paddies, and the crop itself to improve their crop management practices. This way they have increased their crop yields, while at the same time almost eliminating insecticide use with positive benefits in terms of environmental and human health. About 2 million farmers have benefited from this approach.

In Tanzania, problems surrounding the sustainable use of Lake Manyara, a large freshwater lake, arose following increased usage in recent decades. The formation of the Lake Manyara Biosphere Reserve to combine both conservation of the Lake and surrounding high value forests with sustainable use of the wetland area and low- input agriculture has brought together key users to set management goals. The Biosphere Reserve has fostered studies for the sustainable management of the wetlands, including monitoring the ground water and the chemistry of the escarpment water source.

Clayoquot Sound on the western coast of Vancouver Island, Canada, encompasses forests and marine and coastal systems. The establishment of adaptive management to implement the ecosystem approach at the local level is currently under development with the involvement of indigenous communities, with a view to managers who maintain or increase the area of forest within their properties.

In different parts of the world, farmers are raising crops within mixed ecosystems. In Mexico, they are growing "shade coffee," putting coffee trees in a mixed tropical forest rather than in monoculture plantations that reduce biodiversity. These farmers then rely entirely on natural predators common to an intact ecosystem rather than on chemical pesticides.

Tourists, attracted in large numbers by the spectacular beauty of marine and coastal diversity of the Soufrie area of St. Lucia, had a negative impact on the age-old and thriving fishing industry. In 1992, several institutions joined with fishers and other groups with an interest in conservation and sustainable management of the resources and, together, established the Soufrie Marine Management Area. Within this framework, problems are dealt with on a participatory basis with the involvement of all stakeholders.

Through weekly "farmer field schools," rice farmers in several Asian countries have developed their understanding of the functioning of the tropical rice ecosystem including the interactions between insect pests of rice, their natural enemies, fish farmed

Sustainable use of Biodiversity

The conservation of each country's biological diversity can be achieved in various ways. "In-situ" conservation focuses on conserving genes, species, and ecosystems in their natural surroundings, for example by establishing protected areas, rehabilitating degraded ecosystems, and adopting legislation to protect threatened species. "Ex-situ" conservation uses zoos, botanical gardens and gene banks to conserve species.

Promoting the sustainable use of biodiversity will be of growing importance for maintaining biodiversity in the years and decades to come. Under the Convention, the "ecosystem approach to the conservation and sustainable use of biodiversity" is being used as a framework for action, in which all the goods and services provided by the biodiversity in ecosystems are considered. The Convention is promoting activities to ensure that everyone benefits from such goods and services in an equitable way.

Reporting

Each government that joins the Convention is to report

on what it has done to implement the accord, and how effective this is in meeting the objectives of the Convention. These reports are submitted to the Conference of the Parties (CoP), the governing body that brings together all countries that have ratified the Convention. The reports can be viewed by the citizens of all nations.

The Convention secretariat works with national governments to help strengthen reporting and to make the reports of various countries more consistent and comparable, so that the world community can get a clearer picture of the big Trends. Part of that work involves developing indicators for measuring trends in biodiversity, particularly the effects of human actions and decisions on the conservation and sustainable use of biodiversity. The national reports, particularly when seen together, are one of the key tools for tracking progress in meeting the Convention's objectives.

Thematic Programmes

The Convention's members regularly share ideas on best practices and policies for the conservation and sustainable use of biodiversity with an ecosystem approach. They look at how to deal with biodiversity concerns during development planning, how to promote transboundary cooperation, and how to involve indigenous peoples and local communities in ecosystem management.

The Conference of the Parties has launched a number of thematic programmes covering the biodiversity of inland waters, forests, marine and coastal areas, drylands, and agricultural lands. Cross-cutting issues are also addressed on matters such as the control of alien invasive species, strengthening the capacity of member countries in taxonomy, and the development of indicators of biodiversity loss.

Genetic Resources

An important part of the biodiversity debate involves access to and sharing of the benefits arising out of the commercial and other utilisation of genetic material, such as pharmaceutical products. Most of the world's biodiversity is found in developing countries, which consider it a resource for fueling their economic and social development. Historically, plant genetic resources were collected for commercial use outside their region of origin or as inputs in plant breeding. Foreign bioprospectors have searched for natural substances to develop new commercial products, such drugs. Often, the products would be sold and protected by patents or other intellectual property rights, without fair benefits to the source countries.

The treaty recognises national sovereignty over all genetic resources, and provides that access to valuable biological resources be carried out on "mutually agreed terms" and subject to the "prior informed consent" of the country of origin. When a microorganism, plant, or animal is used for a commercial application, the country from which it came has the right to benefit. Such benefits can include cash, samples of what is collected, the participation or training of national researchers, the transfer of biotechnology equipment and know-how, and shares of any profits from the use of the resources.

Work has begun to translate this concept into reality and there are already examples of benefit-sharing arrangements. At least a dozen countries have established controls over access to their genetic resources, and an equal number of nations are developing such controls.

Financial and Technical Support

When the Convention was adopted, developing countries emphasised that their ability to take national actions to achieve global biodiversity benefits would depend on financial and technical assistance. Thus, bilateral and multilateral support for capacity building and for investing in projects and programmes is essential for enabling developing countries to meet the Convention's objectives.

Traditional Knowledge and Practices

The Convention also recognises the close and traditional dependence of indigenous and local communities on biological resources and the need to ensure that these communities share in the benefits arising from the use of their traditional knowledge and practices relating to the conservation and sustainable use of biodiversity. Member governments have undertaken "to respect, preserve and maintain" such knowledge and practices, to promote their wider application with the approval and involvement of the communities concerned, and to encourage the equitable sharing of the benefits derived from their utilisation.

Convention: the Global Environment Facility (GEF). GEF projects, supported by the United Nations Environment Programme (UNEP), the United Nations Development Programme (UNDP) and the World Bank, help forge international cooperation and finance actions to address four critical threats to the global environment: biodiversity loss, climate change, depletion of the ozone layer and degradation of international waters. By the end of 1999, the GEF had contributed nearly $ 1 billion for biodiversity projects in more than 120 countries.

Biosafety Protocol

Since the domestication of the first crops and farm animals, we have altered their genetic makeup through selective breeding and cross-fertilisation. The results have been greater agricultural productivity and improved human nutrition. In recent years, advances in biotechnology techniques have enabled us to cross the species barrier by transferring genes from one species to another. We now have transgenic plants, such as tomatoes and strawberries that have been modified using a gene from a cold water fish to protect the plants from frost.

Some varieties of potato and corn have received genes from a bacterium that enables them to produce their own insecticide, thus reducing the need to spray chemical insecticides. Other plants have been modified to tolerate herbicides sprayed to kill weeds. Living Modified Organisms (LMOs) are becoming part of an increasing number of products, including foods and food additives, beverages, drugs, adhesives, and fuels. Agricultural and pharmaceutical LMOs have rapidly become a multi-billion- dollar global industry.

Biotechnology is being promoted as a better way to grow crops and produce medicines, but it has raised concerns about potential side effects on human health and the environment, including risks to biological diversity. In some countries, genetically altered agricultural products have been sold without much debate, while in others, there have been vocal protests against their use, particularly when they are sold without being identified as genetically modified.

In response to these concerns, governments negotiated a subsidiary agreement to the Convention to address the potential risks posed by cross- border trade and accidental releases of LMOs. Adopted in January 2000, the Cartagena Protocol on Biosafety allows governments to signal whether or not they are willing to accept imports of agricultural commodities that include LMOs by communicating their decision to the world community via a Biosafety Clearing House, a mechanism set up to facilitate the exchange of information on and experience with LMOs.

Challenges

Some of the major challenges to implementing the Convention on Biological Diversity and promoting sustainable development are:

— Meeting the increasing demand for biological resources caused by population growth and increased consumption, while considering the long-term consequences of our actions.

— Increasing our capacity to document and understand biodiversity, its value, and threats to it.

— Building adequate expertise and experience in biodiversity planning.

— Improving policies, legislation, guidelines, and fiscal measures for regulating the use of biodiversity.

— Adopting incentives to promote more sustainable forms of biodiversity use.

— Promoting trade rules and practices that foster sustainable use of biodiversity.

— Strengthening coordination within governments, and between governments and stakeholders.

— Securing adequate financial resources for conservation and sustainable use, from both national and international sources.

— Making better use of technology.

— Building political support for the changes necessary to ensure biodiversity conservation and sustainable use.

— Improving education and public awareness about the value of biodiversity. Economic development is essential to meeting human needs and to eliminating the poverty that affects so many people around the world.

The sustainable use of nature is essential for the long-term success of development strategies. A major challenge for the 21st century will be making the conservation and sustainable use of biodiversity a compelling basis for development policies, business decisions, and consumer desires.

The Convention has already accomplished a great deal on the road to sustainable development by transforming the international community's approach to biodiversity.

This progress has been driven by the Convention's inherent strengths of near universal membership, a comprehensive and science-driven mandate, international financial support for national projects, world-class scientific and technological advice, and the political involvement of governments. It has brought together, for the first time, people with very different interests. It offers hope for the future by forging a new deal between governments, economic interests, environmentalists, indigenous peoples and local communities, and the concerned citizen. However, many challenges still lie ahead. After a surge of interest in the wake of the Rio Summit, many observers are disappointed by the slow progress towards sustainable development during the 1990s. Attention to environmental problems was distracted by a series of economic crises, budget deficits, and local and regional conflicts. Despite the promise of Rio, economic growth without adequate environmental safeguards is still the rule rather than the exception.

The Convention on Biological Diversity and its underlying concepts can be difficult to communicate to politicians and to the general public. Nearly a decade after the Convention first acknowledged the lack of information and knowledge regarding biological diversity, it remains an issue that few people understand. There is little public discussion of how to make sustainable use of biodiversity part of economic development.

The greatest crunch in sustainable development decisions is the short-versus the long-term time frame. Sadly, it often still pays to exploit the environment now by harvesting as much as possible as fast as possible because economic rules do little to protect long-term interests. Truly sustainable development requires countries to redefine their policies on land use, food, water, energy, employment, development, conservation, economics, and trade.

Biodiversity protection and sustainable use requires the participation of ministries responsible for such areas as agriculture, forestry, fisheries, energy, tourism, trade and finance. The challenge facing governments, businesses, and citizens is to forge transition strategies leading to long-term sustainable development. It means negotiating trade-offs even as people are clamoring for more land and businesses are pressing for concessions to expand their harvests.

The transition to sustainable development requires a shift in public attitudes as to what is an acceptable use of nature. This can only happen if people have the right information, skills, and organisations for understanding and dealing with biodiversity issues.

Governments and the business community need to invest in staff and training, and they need to support organisations, including scientific bodies, that can deal with and advise on biodiversity issues. We also need a long-term process of public education to bring about changes in behaviour and lifestyles, and to prepare societies for the changes needed for sustainability.

10

NATURAL DISASTERS

A natural disaster is the consequence of a natural hazard which moves from potential in to an active phase, and as a result affects human activities. Human vulnerability, exacerbated by the lack of planning or lack of appropriate emergency management, leads to financial, structural, and human losses.

The resulting loss depends on the capacity of the population to support or resist the disaster, their resilience. This understanding is concentrated in the formulation: "disasters occur when hazards meet vulnerability".

Classification of disasters is in fact a classification of hazards. There are events, agents, which have the potentiality of doing harm, that is, hazards. Hazard is a source of danger. It is an unknown and unpredictable phenomenon that causes an event to result one way rather than another.

Even if these hazards materialize one has not a disaster if the community is not vulnerable; that is, if it has the capacity to respond or adjust. Hazards have the potential to cause disasters. Hazards cause disasters only when they meet with vulnerable people: when affecting (a) human life, (b) property, and (c) human activities.

IMPACT OF NATURAL HAZARDS

A Natural Hazard is an unexpected or uncontrollable natural event of unusual magnitude that threatens the activities of people or people themselves. A Natural Disaster is a natural hazard event that actually resulted in widespread destruction of property or caused injury and/or death. Natural hazards are usually classified based on where they occur on the Earth.

Atmospheric hazards are most often weather-related events, while geologic hazards happen on or within the Earth's surface. However, it is important to understand that atmospheric hazards can trigger geologic hazards (such as a thunderstorm producing flooding), and geologic hazards can trigger atmospheric hazards (such as a volcanic eruption producing thunderstorms).

In every person's lifetime, at least one natural hazard will likely have some impact on their life. On average, about 150,000 people are killed worldwide by natural hazards each year. Unfortunately, the economic cost of natural hazards is increasing dramatically. For example, In the 1990's, the average cost of all natural hazards in the United States doubled from $25 billion to $50 billion per year.

Globally, losses from natural hazards are increasing partially due to the exponential increase in human population. This unprecedented number of people has reduced resource availability and forced people to live in marginal areas where natural hazards occur more frequently and often with greater severity. In many developed countries, loss of life is low but economic costs are extremely high. In developing countries, loss of life tends to be very high while economic costs are usually lower.

Besides the direct impacts of natural hazards, such as floodwaters destroying a house, there are usually many indirect impacts, such as floodwaters spreading disease for weeks after the initial destruction. Although these indirect impacts may be less obvious initially, they are often more costly and can add years on to the recovery time from a disaster. As people who live in communities that have been devastated by a natural hazard will often say, there is no such thing as a complete recovery—natural disasters can forever change the way of living.

It is important to be aware of natural hazards because human activities can increase how often a natural hazard occurs and how severe a natural hazard can become. For example, building on top of an unstable slope adds weight to the slope and will increase the probability of the slope collapsing. Understanding when, where, why, and how natural hazards occur is the first step in minimizing their impacts on our lives.

A natural hazard is a situation which has the potential to create an event that has an effect on people. They result from natural processes in the environment and some natural hazards are related - earthquakes can result in tsunamis, drought can lead directly to famine and disease, and so on.

Geological Hazards

Avalanche

Avalanche on the backside (East) of Mt. Timpanogos, Utah at Aspen Grove trail An avalanche is a geophysical hazard involving a slide of a large snow (or rock) mass down a mountain side, caused when a buildup of snow is released down a slope, it is one of the major dangers faced in the mountains in winter. An avalanche is an example of a gravity current consisting of granular material. In an avalanche, lots of material or mixtures of different types of material fall or slide rapidly under the force of gravity. Avalanches are often classified by what they are made of. Notable avalanches include:

1. The 1970 Ancash Earthquake
2. The 1954 Blons Avalanches
3. The 1999 Galtur Avalanche
4. The 2002 Kolka-Karmadon Rock Ice Slide
5. The 1910 Wellington Avalanche

Earthquake

An earthquake is a phenomenon that results from a sudden release of stored energy that radiates seismic waves. At the Earth's surface, earthquakes may manifest themselves by a shaking or displacement of the ground and sometimes tsunamis. 90 per cent of all earthquakes - and 81 per cent of the largest—occur around the 40,000 km long Pacific Ring of Fire, which roughly bounds the Pacific Plate. Many earthquakes happen each day, few of which are large enough to cause significant damage. Some of the most significant earthquakes in recent times include:

1. The 2004 Indian Ocean earthquake, the second largest earthquake in recorded history, registering a moment magnitude of 9.3. The huge tsunamis triggered by this earthquake cost the lives of at least 229,000 people.
2. The 7.6-7.7 2005 Kashmir earthquake, which cost 79,000 lives in Pakistan.
3. The 7.7 magnitude July 2006 Java earthquake, which also triggered tsunamis.

Lahar

A Lahar is a type of natural disaster closely related to a volcanic eruption, and involves a large amount of material, including mud, rock, and ash sliding down the side of the volcano at a rapid pace.

These flows can destroy entire towns in seconds and kill thousands of people. The Tangiwai disaster is an excellent example, as is the one which killed an estimated 23,000 people in Armero, Colombia, during the 1985 eruption of Nevado del Ruiz.

Landslides

A landslide is a disaster closely related to an avalanche, but instead of occurring with snow, it occurs involving actual elements of the ground, including rocks, trees, parts of houses, and anything else which may happen to be swept up.

Landslides can be caused by earthquakes, volcanic eruptions, or general instability in the surrounding land. Mudslides, or mud flows, are a special case of landslides, in which heavy rainfall causes loose soil on steep terrain to collapse and slide downwards; these occur with some regularity in parts of California after periods of heavy rain.

Sinkholes

A localised depression in the surface topography, usually caused by the collapse of a subterranean structure, such as a cave. Although rare, large sinkholes that develop suddenly in populated areas can lead to the collapse of buildings and other structures.

Volcanic Eruption

Pu'u 'O'oA volcanic eruption is the point in which a volcano is active and releases its power, and the eruptions come in many forms. They range from daily small eruptions which occur in places like Kilauea in

Hawaii, or extremely infrequent supervolcano eruptions (where the volcano expels at least 1,000 cubic kilometres of material) in places like Lake Taupo, 26,500 years ago, or Yellowstone Caldera, which has the potenetial to become a supervolcano in the near geological future. Some eruptions form pyroclastic flows, which are high-temperature clouds of ash and steam that can trial down mountainsides at speed exceeding an airliner.

Hydrological Hazards

Flood Disaster

Floods are a natural behaviour of streams. Size of an unmodified stream channel is directly related to the quantity of water that it usually carries (more frequent, moderate flows). Most of the time river stage is below the channel banks. Times of higher discharge overflow banks leading to floods. In Bank-full Stage, water fills the channel to the level of the bank top and the Bank-full Discharge is the water discharge when the water level is at the tops of the stream banks. Recurrence interval for bank-full discharge can have varying periods and higher discharge flood events occur much less frequently. Recent events especially hurricanes of the Caribbean region and notably IVAN, have caused greater damages to property, lives and housing. Floods were also caused by the incessant and torrential rains across the Americas.

Major Flood Disasters:

1. Hwnag Ho River, China 1887: estimated 900,000 fatalities.
2. Johnstown, Pennsylvania 1889 Dam failed: 2,200 fatalities.
3. Yangtze, China 1911: estimated 100,000 fatalities.
4. Los Angeles, CA 1928 St. Francis Dam fails.

5. Yangtze, China 1931: estimated 200,000 fatalities.
6. Huang He, China 1938: estimated 900,000 fatalities.

Floods in India

India is prone to floods, just as any other country in the world. What is worse however is that when rivers in one part are in spate, the land in other parts could be parched. Floods and droughts always occur at one and the same time, making life difficult for the people on both the counts. Floods are a common, everyday occurrence and in floodplains more so than anywhere. The Indian Meteorology Department uses a simple classification of floods, and defines that rainfall in excess of 75 per cent causes very severe floods, 50 per cent severe floods and 25 per cent moderate floods.

The State of Assam, for example, located in the northeast region of India, has become a multi-disaster prone area. Due to deforestation in the upper catchment areas of the rivers and the lack of proper maintenance of dams and protective embankments, the region is becoming more and more vulnerable to flooding.

In the last two years, the monsoon floods have become a nightmare to thousands of resource-poor people living at the side of the mighty Brahmaputra river. Every year, the river is becoming wider and wider, consuming vast areas of fertile land and human habitations. With the advent of the monsoon in June 2004, for example, there was heavy rain in the entire region, including Bhutan.

When the Kuriso Dam in Bhutan was in danger of overflowing in early July the authorities released unexpectedly large amounts of water, which caused an increase in the level of the Brahmaputra River. In early July, it again started to rain heavily in this area and water levels in the rivers started rising again. From July 10, 2004 onwards, the rivers started overflowing their banks and more and more areas became inundated resulting in a major flood. At several places, breaches developed in the embankments and torrents of water swept through villages and washed away a large number of houses. Twenty three districts have been severely affected by the floods. Goalpara and Dhubri districts are among the worst affected.

Flood Hazard Categories

It is necessary to divide the floodplain into flood hazard categories that reflect the flood behaviour across the floodplain. Commonwealth Scientific and Industrial Research Organisation (CSIRO) refers to the degree of flood hazard as being a function of:

— the size (magnitude) of flooding;

— depth and velocity (speed of flowing water);

— rate of floodwater rise;

— duration of flooding;

— evacuation problems;

— effective flood access;

— size of population at risk;

— land use;

— flood awareness/readiness;

— effective flood warning time.

There are four degrees of flood hazard: low, medium, high and extreme. The categorisation of the floodplain is largely qualitative using the above factors. For example, medium hazard is where adults could wade safely, but children and elderly may have difficulty, evacuation is possible by a sedan, there is ample time for flood warning and evacuation and evacuation routes remain trafficable for at least twice as long for the required evacuation time.

A key factor in the case of evacuation from an area is the water depth and the velocity along the evacuation route; that is, the stability of pedestrians wading through flood waters or vehicles driving along flooded roads. There are some estimation procedures available for stability estimation, but further research is required across a broader range of conditions.

Recommended Approach

In considering the application of flood related issues to the specific flood characteristics of the lower Johnstone River floodplain, it is noted that:

1. Duration of flooding is universally long (in the order of days) across the floodplain;
2. Warning times can be short (~ 6 hrs);
3. Rates of floodwater rise are reasonably fast; and
4. Flood-awareness is generally high and does not vary significantly across the floodplain.

The four parameters are not significantly variable across the floodplain to warrant specific treatment and are therefore not used to define variations in the flood hazard, but should be included in development control measures. The flood hazard is therefore defined on the remaining, varying characteristics of:

1. The size of the flood;
2. Depth and velocity of floodwaters; and
3. Evacuation and access.

Flood Warning

In the Johnstone region, for example, many people come from families that have resided in the area for several generations. In most cases, these people have either experienced a flood or have heard first hand accounts of floods from family members or friends. Therefore, they are likely to have a high level of flood awareness. However, these people may not be aware that there may be larger floods than those events that they have experienced or heard of. In addition, there are a significant number of new rural and urban residents in the region who may not have the same level of flood awareness. In some instances, these people:

1. have not experienced a flood in the area;
2. have not heard first hand accounts of previous floods;
3. live in houses that are not near the river, but are actually in the floodplain and are subject to flooding; and/or
4. are not likely to take flood warnings seriously.

Limnic Eruption

Also referred to as a 'lake overturn', a limnic eruption is a rare type of natural disaster in which CO_2 suddenly erupts from deep lake water, posing the threat of suffocating wildlife, livestock and humans. Such an eruption may also cause tsunamis in the lake as the rising CO_2 displaces water. Scientists believe landslides, volcanic activity, or explosions can trigger such an eruption.

To date, only two limnic eruptions have been observed and recorded:

1. In 1984, in Cameroon, a limnic eruption in Lake Monoun caused the deaths of 37 nearby residents.
2. At nearby Lake Nyos in 1986 a much larger eruption killed between 1,700 and 1,800 people by asphyxiation.

Maelstrom

A large tidal whirlpool. The largest known maelstrom is Moskstraumen off the Lofoten islands in Norway. Powerful whirlpools have killed unlucky seafarers, but their power tends to be exaggerated in fiction. Maelstroms can reach speeds of 20-40 km/h. A seiche is a standing wave in an enclosed or partially enclosed body of water. Seiches and seiche-related phenomena have been observed on lakes, reservoirs, bays and seas. The key requirement for formation of a seiche is that the body of water be at least partially bounded, allowing natural phenomena to form a standing wave.

Tsunami

A tsunami is a wave of water caused by the displacement of a body of water. Tsunami can be caused by undersea earthquakes as in the 2004 Indian Ocean Earthquake, or by landslides such as the one which occurred at Lituya Bay, Alaska. Meteotsunamis are caused by meteorological phenomena. A megatsunami is an informal term used to describe very large tsunamis.

They are a highly local effect, either occurring on shores extremely close to the origin of a tsunami, or in deep, narrow inlets. The largest waves are caused by very large landslides, such as a collapsing island, into a body of water. The highest Tsunami ever recorded was estimated to be of 524 m (1742 ft.) vertical run-up on July 10, 1958, in Lituya Bay, Alaska.

Climatic

Blizzard

A severe winter storm condition characterised by low temperatures, strong winds, and heavy blowing snow. Significant blizzards in the United States include:

— The Great Blizzard of 1888.

— The Schoolhouse Blizzard earlier the same year.

— The Armistice Day Blizzard in 1940.

— The Storm of the Century in 1993.

Drought

An abnormally dry period when there is not enough water to support agricultural, urban or environmental water needs. Extended droughts can result in deaths by starvation or disease, and can result in wildfires. Well-known historical droughts include:

— 1900 India, killing between 250,000 and 3.25 million.

— 1921-22, Soviet Union, in which over 5 million perished from starvation due to drought.

— 1928-30, northwest China, resulting in over 3 million deaths by famine.

— 1936 and 1941, Sichuan Province, China, resulting in 5 million and 2.5 million deaths respectively.

As of 2006, Western Australia, New South Wales, Victoria and Queensland (all states of Australia) have been under drought conditions for five to ten years.

The drought is beginning to affect urban population for the first time. Also in 2006, Sichuan Province, China experienced its worst drought in modern times, with nearly 8 million people and over 7 million cattle facing water shortages. Scientists warn that global warming may result in more extensive drought in coming years.

Numerous Hailstorm

A hailstorm is a natural disaster where a thunderstorm produces numerous hailstones which damage the location in which they fall. Hailstorms can be especially devastating to farm fields, ruining crops and damaging equipment. A particularly damaging hailstorm hit Munich, Germany on August 31, 1986, felling thousands of trees and causing millions of dollars in insurance claims.

Heat Wave

A heat wave is a disaster characterised by heat which is considered extreme and unusual in the area in which it occurs. Heat waves are rare and require specific combinations of weather events to take place, and may include temperature inversions, katabatic winds, or other phenomena.

The worst heat wave in recent history was the European Heat Wave of 2003. There is also the potential for longer term events causing global warming, including stadial events, or through human induced climatic warming.

Hurricanes, Tropical Cyclones, and Typhoons

Hurricane, tropical cyclone, and typhoon are different names for the same phenomenon: a cyclonic storm system that forms over the oceans. It is caused by evaporated water that comes off of the ocean and becomes a storm. The Coriolis Effect causes the storms to spin, and a hurricane is declared when this spinning mass of storms attains a wind speed greater than 74 mph.

Hurricane is used for these phenomena in the Atlantic and eastern Pacific Oceans, tropical cyclone in the Indian, typhoon in the western Pacific. The deadliest hurricane ever was the 1970 Bhola cyclone; the deadliest Atlantic hurricane was the Great Hurricane of 1780, which devastated Martinique, St. Eustatius and Barbados.

Ice Age

An ice age is a geologic period, but could also be viewed in the light of a catastrophic natural disaster, since in an ice age, the climate all over the world would change and places which were once considered habitable would then be too cold to permanently inhabit. A side effect of an ice age could possibly be a famine, caused by a worldwide drought.

Ice Storm

An ice storm is a particular weather event in which precipitation falls as ice, due to atmosphere conditions.

Tornado

A tornado is a natural disaster resulting from a thunderstorm. Tornadoes are violent, rotating columns of air which can blow at speeds between 50 and 300 mph, and possibly higher. Tornadoes can occur one at a time, or can occur in large tornado outbreaks along squall lines or in other large areas of thunderstorm development. Waterspouts are tornadoes occurring over tropical waters in light rain conditions.

11

ENVIRONMENTAL POLLUTION

Today environmental pollution is one of the most serious problems facing mankind. Air, water, and soil are essential to the survival of all living creatures on planet earth, but unfortunately they are harmed by pollution. Air polluted beyond tolerable levels can cause illness, and even death. Polluted water kills fish and other marine life and is dangerous to human health. Pollution of soil decreases the amount of land available for growing crops, vegetables and fruits.

The naturally beautiful world is becoming ugly by environmental pollution. Everyone is now conscious and wants to reduce pollution. But the pollution problem is both serious and also complicated. It is complicated because much of pollution is caused by things that benefit mankind. For example, automobile exhaust causes a large percentage of all air pollution. However, the automobile provides transportation for millions of people all over the world.

Discharge from factories is responsible for much of the material that pollutes air and water, but factories provide jobs and produce goods that people want. The use of fertiliser or pesticide can ruin soil, but fertilisers and pesticides are important aids to the growing of crops, fruits and vegetables. In order to end or greatly reduce pollution immediately, people must stop using many things that benefit them. Unfortunately, most people do not want to do that. But pollution can be reduced in a planned manner in several ways. Scientists and engineers can work together to find methods to decrease the amount of pollution that such things as automobiles and factories cause.

Governments should pass and enforce laws that require manufacturers and individuals to stop, or reduce on certain polluting activities. And most important of all is individuals and people's organisations must work to persuade their representatives in government, and also persuade manufacturers, to take action toward reducing pollution.

People have always polluted their environment or caused some environmental pollution. Since prehistoric times, human beings have put wastes in water and generated smoke by burning fuel. However, in early civilisation people did not live crowded together, and they had no pollution-causing machines. Therefore, pollution was negligible and it was spread out over large areas. During ancient times, pollution problems first arose when large numbers of people started to live together in cities. With the growth of the cities, pollution also increased proportionately with them. However, the environmental problems were not serious or widespread until eighteenth century and early part of the nineteenth century. This period in history is called the Industrial Revolution, which began in England and spread to other European countries and the United States.

The notable feature of the Industrial Revolution was the development of factories and overcrowding with factory workers in cities. At that time coal was the prime energy fuel to power most of the factories and to heat most of the homes in the cities. Because of the burning of coal, the air over such industrial cities as London became filled with huge amounts of smoke and soot containing sulphur dioxide and nitrogen dioxide. An additional problem was poor sanitation facilities, which allowed raw sewage to get into water supplies in some cities, which sometimes happens in the developing countries even now. The polluted water causes typhoid fever and other diseases.

In the early part of the twentieth century, air pollution in industrial cities in the United States became a particularly serious problem. During the post economic depression period, smoke and smote emanating from factories, steel mills, power plants, railroads, and heating plants filled the air over many Eastern and Midwestern cities. It was reported that in some cities such as Pittsburgh and St. Louis, the air pollution was so thick that the motorists had to use headlights to see during the daylight hours.

After the end of the Second World War, air pollution from coal burning has been greatly diminished in most parts of the world. Cleaner fuels such as oil and natural gas are being used in nearly all railroads and many industries and home heating plants now. Those factories, which use coal, have taken steps to control the pollution from their furnaces. In many countries, epidemics from disease-carrying germs in city water supplies are no longer a major problem. Sanitary Engineering is advanced, so that cities now treat their water and keep it as free from germs as possible.

Inspite of these technological advancements, environmental pollution has become increasingly serious and widespread. Technological improvements have helped cause this increase. In addition, the population growth in urban areas is tremendously increasing. A rule of thumb is more the people more is the waste produced of every kind. Public awareness of environmental pollution has grown considerably. Dramatic environmental tragedies that occurred during the 1960's and the 1970's have pointed up the seriousness of the pollution problem. These include the oil spills that ruined the beautiful beaches and killed wildlife and waterways that had become nearly dead from solid wastes.

During the late 1970's the inhabitants of the city of Seveso in Italy were threatened after an explosion at a local chemical factory that released a poisonous and carcinogenic gas called dioxin. Chemical wastes that leaked from a former disposal site called Love Canal, in Niagara Falls, New York, resulted in many families moving out of their homes located in the area. Chemicals, such as polychlorinated biphenyls (PCB's) were discharged in the wastewater of certain industrial plants, found their way into the bodies of fish and other animals in various parts of the United States.

PUBLIC CONCERN FOR ENVIRONMENTAL POLLUTION

People have always polluted the environment. But for large part of history, pollution was not a major problem. Most of the people lived in uncrowded rural areas and the pollutants (waste products) they produced were scattered widely. At that time pollution-causing machines were not invented, such as industrial plants or motor vehicles.

Industrialisation and the development of crowded cities in the 1700's and the 1800's made pollution a major problem to man, plants and animals. Factories and human beings in these urban cities released huge amounts of pollutants into small areas. With the advent of automobiles and the rapid development of urban areas made pollution steadily worse. By the middle of the twentieth century, pollution had affected the water in every major lake and river and the air over every major city In the United States and other Industrial nations. Since man landed on the moon, millions of people have become alarmed by the dangers of pollution.

The United Nations and a large number of countries are joining hands and working together to reduce pollution. There are several kinds of environmental pollution. They are air pollution, soil pollution, water pollution, and pollution caused by solid wastes, noise, and radiation. All parts of the environment are closely related to one another, ecologically. Ecology is the study of the relationships among living things, and between living things and other parts of the environment. Because of the ecological relationships, a kind of pollution that chiefly harms one part of the environment may also affect others.

INFLUENCE OF AIR POLLUTION

Air pollution turns clear, odorless air into hazy, smelly air that harms health, kills plants, and damages property. Human beings cause air pollution by pouring hundreds of millions of tons of gases and particulates into the atmosphere each year. Particulates are tiny particles of solid or liquid matter. Smog is one of the most common forms of air pollution. Combustion (burning) is responsible for most of the air pollution. The burning of gasoline to power by motor vehicles and the burning of coal to heat buildings and help manufacture products are examples of such processes. In a combustion process, each time a fuel is burned, some type of pollutant is released into the air.

The pollutants vary from small amounts of colourless poison gas to clouds of thick black smoke. Weather conditions can help to decrease the amount of pollutants in the air. Wind is responsible for scattering the pollutants, and rain and snow wash them into the ground. Unfortunately, in many areas, pollutants are discharged into the air faster than weather conditions can dispose of them. As we are witnessing everyday, in all crowded cities throughout the world, for example, millions or thousands of automobiles, factories, and furnaces are adding tons of pollutants into a small area of the atmosphere each day.

Sometimes, weather conditions cause pollutants to build up over an area instead of clearing them away. One such weather condition -called thermal inversion-occurs when a layer of warm air settles over a layer of cooler air, which is near to the ground. The warm air holds down the cool air and prevents the pollutants from rising and scattering. A serious pollution problem occurs when a thermal inversion prevails over a city that is pouring tons of pollutants into the air.

IMPORTANCE OF WATER POLLUTION

As a result of water pollution the amount of pure, fresh water that is available is reduced for such necessities as drinking and cleaning, and for such activities as swimming and fishing. Water is affected by pollutants that come mainly from industries, farms, and sewerage systems. It is well known that factories dump huge amounts of waste products into bodies of water each year. The wastes that are dumped into water reservoirs include chemicals, wastes from animal and plant matter, and hundreds of other substances. Wastes from farms include animal wastes, fertilisers, and pesticides.

A large amount of these materials drain off from farm fields into nearby bodies of water. Sewerage systems carry wastes from homes, offices, hospitals, and factories into water. Almost every city has a waste treatment plant that removes some of the most harmful wastes from sewage. This does not remove completely materials that harm the water.

Natural cycles work to absorb small amounts of wastes in reservoirs of water. During a natural cycle, wastes are turned into useful or at least harmless substances. Aerobic bacteria use oxygen to breakdown natural wastes such as dead fish and decay them down into chemicals, including nitrates, phosphates, and carbon dioxide. These chemicals, called nutrients, are used as food by algae (simple organisms) and other green plants in the water. Microscopic animals called zooplankton eat algae as food. Small fish such as minnows eats zooplankton. The small fish in turn, are eaten by larger fish.

The larger fish eventually die and are broken down by bacteria. The cycle then begins again. This same natural cycles work on wastes discharged into water by people. Chemicals are broken down by bacteria. Bacteria also break down other wastes and turn them into nutrients, or else into substances that will not harm fish or sea plants. On the other hand, if too much waste matter is poured into the water, the whole cycle will begin to break down, and the water starts to become dirtier and dirtier. The bacterium that works to decay the wastes consumes too much oxygen during the decaying process.

As a consequence of this, less oxygen is available for the animals and plants in the water. Due to lack of oxygen the animals and plants die, adding even more wastes to the water. In the end, the water's entire oxygen supply is used up and, without oxygen, anaerobic bacteria, rather than aerobic bacteria, decay wastes. The disadvantage of the anaerobic decaying process is that it causes wastes to give off smelly gases.

A process called nutrient enrichment, or eutrophication, takes place by nutrients present in the water. People also add nutrients to water, such as nitrates from agricultural fertilisers and phosphates from detergents in sewage; greatly increase the growth of algae in water. Growth of larger amounts of algae also results in the death of large amounts of algae. The dead algae become wastes, and, as they decay they use up the water's oxygen supply.

When heated water is added to a body of water, it upsets the cycles. Heated water can kill animals and plants, which are accustomed to living at lower temperatures in the water. Heated water reduces the amount of oxygen that water can hold. The addition of heated water is called thermal pollution. The source of most heated water is industries and nuclear power plants that use water for cooling.

SOIL POLLUTION

Soil pollution damages the thin layer of fertile soil that covers much of the earth's land and is essential for growing crops, vegetables and fruits. Nature takes thousands of years to form the soil to support food crops, but man can destroy it in a few years. In nature, cycles keep soil fertile. Bacteria and fungi decay these wastes such as animal wastes and dead organisms, breaking them down into nitrates, phosphates, and other nutrients. The nutrients feed growing plants, and when the plants die the cycle begins again. The use of large amounts of fertilisers decreases the ability of bacteria to decay wastes and produce nutrients naturally. Pesticides harm bacteria and other helpful organisms in the soil. Erosion causes much damage to soil. Careless farming and land construction cause erosion.

Production of Solid Waste

The most visible form of pollution is solid wastes. Garbage and trash results in billions of tons of solid material each year. Much of this waste ends up littering roadsides, floating in lakes and streams, and collecting in ugly dumps. Some examples of solid wastes are junked cars, tires, and refrigerators. Stoves, cans and other packaging materials, and scraps of metal and paper.

When solid waste is stored' dumps, it provides home for disease-carrying animals, such as cockroaches and rats. When solid waste is destroyed by burning it, the smoke causes air pollution. Wastes dumped into water results in water pollution. Each American produces about 45 pounds (20 kilograms) of solid waste each day. Production of solid waste is increasing rapidly.

HUMAN HEALTH AND ENVIRONMENTAL POLLUTION

Many studies in people have demonstrated an association between environmental exposure and certain diseases or other health problems. Examples include radon and lung cancer; arsenic and cancer in several organs; lead and nervous system disorders; disease-causing bacteria such as E. coli O157: H7 and gastrointestinal illness and death; and particulate matter and aggravation of heart and respiratory diseases.

To understand the relationship between health and the environment, scientists study a series of events that begins with the release of a pollutant into the environment and may end with the development of disease in a person or a population.

Role of the Environment in Disease

Decades of research have provided the scientific foundation for understanding the role of the environment in disease. For many pollutants, scientists know with some certainty that exposure to these pollutants, at sufficiently high concentrations, can cause a variety of health effects. For other pollutants, where scientific evidence is less conclusive, scientists can only establish an "association" between exposure and health problems.

Some effects on health may be short-term and reversible, such as irritated eyes from smog. Other effects, such as emphysema, heart disease, and cancer are chronic or even fatal. Some effects may appear shortly after exposure. Others, such as cancer, may require a long lead time before the disease appears.

In many cases, pollution likely is just one of several factors—including diet, exercise, alcohol consumption, and genetic make-up—that influence whether an exposed person will ever become sick. Although exposure to ETS is associated with lung cancer, whether a person will develop cancer from that exposure depends on the amount, frequency, and length of exposure, exposure to other contaminants, and personal characteristics (genes) and behavior (diet and other lifestyle choices).

All these factors can be important in illness and premature death, but they are poorly understood, difficult to quantify, and not routinely tracked or reported. Because of these complexities, it is very difficult to establish causal relationships, and few diseases are known to be exclusively the result of exposure to an environmental pollutant. In many cases, only a small portion of the national incidence of a particular disease is likely to be attributed to a specific environmental factor.

Further complicating the picture is the fact that several segments of the population may be at higher risk for damage or disease from environmental pollutants. Potentially sensitive groups include children; older Americans; people with existing health problems such as diabetes, respiratory disease, or heart disease; and persons with compromised immune systems, including those who have HIV/AIDS or are undergoing cancer chemotherapy. Poor or other disadvantaged populations may live in more polluted environments that expose them to higher concentrations of pollutants. Understanding the impacts of pollutants on such sensitive groups is important for those people directly, as well as for the development of protective national health standards and policies.

Children may be more vulnerable to some environmental pollutants than adults for a number of reasons related to their size, growth, and behaviours. Further, children may become ill from exposures that would not affect adults.

Older Americans may also be especially vulnerable to harmful health effects associated with environmental pollutants, in part because some health problems take many years to develop. A long life span may provide the time needed for occupational or cumulative environmental exposures to induce illness or disease. Also, because of medical advances, many older Americans may be living with health conditions that previously shortened life spans. And, older Americans may have preexisting conditions—such as heart ailments, diabetes, or respiratory disease—that reduce their tolerance to pollutants.

Even relatively healthy older people may, merely as a result of age, have a diminished capacity to fight infections, pollution, or other causes of stress to their systems that might have posed little risk when they were younger. Harmful substances may be processed and eliminated from the body more slowly in older people, which can prolong exposure to those substances and increase susceptibility to associated health problems. Older people are also more likely to become dehydrated and experience other serious consequences of gastrointestinal disease.

Sorting out the role of all these risk factors—including the environment—and their interactions is a major challenge of scientific research.

Health Effects of Air Pollution

Several outdoor air pollutants are associated with harmful health effects. These include the six "criteria" pollutants—particulate matter, ground-level ozone, nitrogen dioxide, carbon monoxide, sulphur dioxide, and lead—for which EPA has established standards to protect human health, including the health of sensitive populations such as asthmatics, children, and the elderly. The burning of fossil fuels is the principal source of these pollutants. Air pollutants can be transported long distances, so they can potentially have effects distant from their source.

Air pollution has been associated with several health problems, including reported symptoms (nose and throat irritation), acute onset or exacerbation of existing disease, and premature deaths. The impact of air pollution on health was underscored in December 1952 when a slow-moving area of high pressure came to a halt over the city of London. Fog developed over the city, and particulate and sulphur pollution began accumulating in the stagnating air mass. Smoke and sulphur dioxide concentrations built up over 3 days. Mortality records showed that deaths increased in a pattern very similar to that of the pollution measurements. An estimated 4,000 extra deaths occurred over a 3- to 4- day period. This represents the first quantitative air pollution exposure data with a link to health.

While the London episode highlighted the hazard of extreme air pollution episodes, it was unclear whether health effects were associated with lower concentrations. By the 1970s, the association between respiratory disease and particulate and/or sulphur oxide air pollution had been well established. Improvements in the measurement of air pollution and health endpoints, plus advances in analytical techniques, have made it possible to quantitatively evaluate air pollution and health.

Particulate air pollution is associated with increased daily mortality in many U.S. communities and other countries. The elderly and those with preexisting diseases are particularly vulnerable. Exposure to ambient particulate matter has also been associated with an increased number of hospital admissions and visits to doctors due to cardiovascular problems and respiratory disease. Some studies show that exposure to particulate matter exacerbates asthma. Long-term exposure to particulate matter has been associated with increased deaths from heart and lung diseases, increased respiratory disease and bronchitis and with decreased lung function in children.

Ozone Exposures

Repeated short-term exposures to ozone may damage children's developing lungs, which may lead to permanent reductions in lung function. Controlled studies in healthy adults have demonstrated ozone-induced lung inflammation, decrements in lung function, and associated respiratory symptoms, such as cough and pain on deep inspiration. Ozone exposures have also been associated with an increased number of hospital admissions and visits to doctors.

Health Indicators

Possible future health indicators for air pollution include death due to respiratory and cardiovascular disease, increased hospital admissions for respiratory and cardiovascular disease, and subtle changes in the cardiovascular system that can increase people's risk of heart attacks and other cardiovascular effects. Use of these indicators is still challenged by limits in the understanding of how much air pollution contributes to the risk of cardiovascular and respiratory disease.

Cause of Waterborne Diseases

In the early 20th century, waterborne diseases such as cholera and typhoid fever were major health threats across the U.S. Deaths due to diarrhoea-like illnesses, including typhoid, cholera, and dysentery, were the third largest cause of death in the nation. For instance, more than 150 in every 100,000 people died from typhoid fever each year.

Around that time, scientists began to understand the cause of these diseases. They had identified the bacteria responsible for most diarrheal deaths and elucidated how these bacteria were transmitted to and among humans. Infected and diseased individuals shed large quantities of microbes in their feces, which flowed into and contaminated major water supplies. This contaminated water was then distributed untreated to communities, which used the water for drinking and other purposes. This created a continuous transmission cycle.

Once treatment of drinking water was initiated to remove pathogens, the number of deaths due to diarrheal diseases dropped dramatically in communities with treated water. Deaths due to typhoid fever were tracked throughout the early 20th century, as drinking water treatment was implemented across the country, providing an indicator of the success of this environmental management strategy.

Drinking water treatment is one of the great public health success stories of the 20th century. Not only did it dramatically and significantly reduce death rates from waterborne disease, it also increased life expectancy and reduced infant mortality.

Today, public health is protected against new and emerging waterborne microbial contaminants by continual improvements to the drinking water treatment process.

This example illustrates how a link was made between gastrointestinal disease (an outcome indicator) and exposure to pathogens in drinking water. Based on this connection, officials were able to take effective action to protect public health. They also were able to use an outcome measure (deaths due to typhoid) to monitor the success of these protective actions.

Today, deaths due to typhoid, cholera, and dysentery are so rare in the U.S. that they cannot serve as indicators to evaluate drinking water management decisions. The actual number of cases of typhoid, cholera, and dysentery are tracked to some extent; however, the reporting of these cases is not federally required.

The waterborne disease outbreak surveillance system is a passive system in that it relies on state health departments to voluntarily report their outbreaks to CDC.

EXPOSURE TO ENVIRONMENTAL POLLUTION

"Exposure" refers to direct human contact with a pollutant (e.g., through breathing contaminated air, drinking contaminated water, or eating contaminated food). Measurements of such exposures can help identify which pollutants may cause health problems and at what levels. They can also provide the basis for determining appropriate actions to limit exposure and associated harmful health effects. For example, these data can enable health officials to respond to a health threat in a specific community.

Environmental Measurements

Historically, human exposure has often been estimated through environmental measurements of ambient pollutant concentrations (e.g., particulate matter in air, bacteria in water or food). However, the presence of a pollutant in the environment does not necessarily mean that anyone is exposed.

Monitoring ambient pollutant levels is critical to measuring exposure for several pollutants, including air pollution, radiation, biological pollutants, and disinfection by-products, which are formed when chlorine is used to treat drinking water. For instance, measurements of concentrations of pollutants in outdoor or indoor air can be coupled with human activity patterns to estimate human exposures. This approach was used to establish national air and water quality standards for many pollutants that protect the U.S. population from harmful health effects.

Personal Monitoring

With personal monitoring, the monitoring device is worn by individuals as they proceed through their normal activities. This approach is most common in workplaces. The radioactivity sensors worn by nuclear power plant workers are one example. Personal monitoring has been used to estimate total human exposures, including exposures from the air people breathe, the water they drink, and the food they eat. One advantage of personal monitoring is that the data provide valuable insights into the sources of the pollutants to which people are actually exposed. A challenge with personal monitoring is ensuring that sufficient sampling is done to be representative of the population being studied.

Biomonitoring

Several environmental pollutants, notably heavy metals and some pesticides, can accumulate in the body over time, often with increasing risk of harm. These pollutants or their breakdown products leave residues in the body that can be measured, usually in the blood or urine. These residues reflect the amount of the pollutant in the environment that actually gets into the body. The approach of measuring pollutant levels in tissue or fluid samples from individual people is called "biomonitoring."

National-scale biomonitoring data can be useful as indicators of the distribution of exposure across the entire population to a variety of pollutants. Also, such data provide an important bridge to understanding the relationships between ambient pollutant concentrations (e.g., in air, water), exposures to these pollutants, and health problems. Biomonitoring data provide exposure information that may help alert physicians, scientists, and health officials to diseases that result from exposure to environmental chemicals. The data are also useful for establishing reference ranges that can be used to identify people with unusually high exposure or the percentage of the population that has pollutant exposures above levels considered to be elevated (e.g., lead).

Health and environmental agencies are using biomonitoring measures and trend data to improve understanding of the relationship between exposure to environmental pollutants and health. Biomonitoring data are already available for metals (e.g., lead, mercury, cadmium), cotinine (a measure of environmental tobacco smoke [ETS]), volatile organic chemicals, organophosphate pesticides, organochlorine pesticides, phthalates, polychlorinated biphenyls (PCBs), dioxin and dioxin-like compounds, and polycyclic aromatic hydrocarbons (PAHs). Future biomonitoring will build the trend data showing whether levels of other pollutants are increasing or decreasing in the population.

Although biomonitoring data are highly useful, they have several limitations as an indicator of exposure. These data do not provide information about how the exposure occurred or the source(s) of exposure, and in some cases, they do not distinguish among different pollutants that may leave identical residues in the body.

CHALLENGES IN HUMAN HEALTH INDICATORS

Human health indicators provide important tools that regulatory agencies can use to identify environmental health problems, develop programmes to reduce the problems, then gauge the success of those efforts. For many other pollutants, major knowledge gaps and challenges remain in linking environmental pollution to health problems. Sorting out the role of the environment, the role of other factors (e.g., genetic make-up, lifestyle choices such as diet and exercise), and the importance of their interactions remains an enormous scientific challenge.

The time between exposure and the development or diagnosis of disease, as well as the problems of tracking a mobile population, further complicate the issue of clarifying connections between exposure and harm to health.

An emerging area of science involves examining the possible combined (additive), synergistic, and cumulative effects of numerous pollutants in the environment. This field of study merits greater development. Finally, not all chemical exposures result in harm to health. With a better understanding of the contribution of environmental factors to the development of disease, EPA will be able to use established health outcome measures—disease trend and exposure data—to enhance environmental management efforts and to assess the effectiveness of those efforts.

Disease registries could be improved to provide valuable assistance in tracking many diseases. Currently, most disease indicators are based on mortality data, which have serious limitations for linking environmental exposures to disease. Data on the number of new cases (incidence) of a disease or the existing cases (prevalence) of a disease in a population can provide better information, but no comprehensive nationwide systems exist for collecting these data.

Occurrence of endemic waterborne disease is grossly underreported. Submission of waterborne disease information to CDC is strictly voluntary, and state-level data pose problems because the list of gastrointestinal diseases that must be reported varies by state. Also, for an outbreak to be detected, many people need to become ill at the same time, and many cases go unreported or are not diagnosed.

Better national-level disease data that could be linked directly with environmental monitoring data would support efforts to establish connections between disease and environmental exposures. For meaningful comparisons, all data sets should have similar timeframes and locations.

Also, national-level efforts would benefit from more data that can be sorted by several relevant factors, such as race (which can help in identifying disparities in health status and outcomes), income, occupation, and residence. Such data can be gathered only through better collaboration between and among environmental and health agencies at all levels, as well as hospitals, clinics, and medical offices. As EPA works to develop environmental indicators that reliably signal trends in exposures and disease, the Agency will also work to improve cooperation with the federal and state agencies that collect relevant information.

Appropriate indicators that address these challenges can help the agencies responsible for monitoring and managing the nation's health to flag and respond to potential problems, such as an upsurge in cases of an environmentally related disease or rising contaminant levels in human tissues. The same indicators might, ideally, show whether pollution control actions are actually reducing the number of people who develop diseases associated with environmental agents. This information will help EPA and other agencies to enhance priority-setting to best protect the health of the nation's people.

12

LEGAL FRAMEWORK FOR ENVIRONMENTAL PROTECTION

The regulatory and institutional decision-making framework for environmental management and protection in India, both at the National and State level, is embodied in nine major acts of the Indian Parliament. Among those, the ones applicable to Water Resource development projects are:

1. Water (Prevention and Control of Pollution) Act of 1974.
2. Air (Prevention and Control of Pollution) Act of 1977.
3. Environment (Protection) Act of 1986, the Forest (Conservation) Act of 1980, amended in 1988.
4. Public Liability Insurance Act of 1991.
5. Environmental Impact Assessment of Development Projects, of 1994.

The Ministry of Environment and Forests (MoEF) is the nodal agency in the administrative structure of the Central Government for the planning, promotion, co-ordination and overseeing the implementation of environmental legislation and programmes. Regulatory functions like Environment Clearance, Forest clearance are part of the mandate of this agency.

CONSTITUTIONAL PROVISIONS

Water Resources

Under the Constitution of India, which came into force in 1950, 'Water' is primarily a State subject and the Union comes in only in the case of interstate river waters. Entry 17, List II, i.e. State List in 7th Schedule of the Constitution states, "Water, that is to say water supplies, irrigation and canals, drainage and embankments, water storage and water power subject to the provisions of Entry 56 of List I". States are thus free to enact "water" laws and frame policies in accordance with this provision. Entry 56 of List I (Union list) referred to above states, "Regulation and development of interstate rivers and river valleys to the extent to which such regulation and development under the control of the Union, is declared by Parliament by law to be expedient in the public interest".

Under Article 262 of the Constitution, Parliament may, by law (1) provide for the adjudication on any dispute or complaint with respect to the use, distribution or control of the waters of, or in, any inter-state river or river valley and (2) "that neither the Supreme Court nor any other court shall exercise jurisdiction in respect of any such dispute or complaint" as referred to in (1).

Natural Environment

The first constitutional provisions related to environment were made in the Forty-Second Amendment to the Indian Constitution. This amendment was passed in response to India being party to the Stockholm Declaration adopted by the International Conference on Human Environment in 1972. The Forty-Second Amendment inserted Article 48-A into the Directive Principles of State Policy in Chapter IV of the Constitution. This declared the State's responsibility to protect and improve the environment and safeguard the forests and wildlife of the country. Another provision, inserted in Article 51-A (g), stipulated the duty of every citizen to "protect and improve the natural environment including forests, lakes, rivers and wildlife and to have compassion for living creatures." These amendments imposed an obligation on the Government and the courts to protect the environment for the people and the nation.

Article 32 and Article 226 of the constitution confer power on the Supreme Court and the High Courts to issue writs in the nature of habeas corpus, mandamus, certiorari and prohibition. These Articles can be used to show that river pollution implicates a public authority that has been vested with the responsibility to prevent pollution but is not executing its powers. These Articles are emancipatory provisions that allow citizens to challenge the structures of domination from within the constitutional framework.

WATER POLICY FRAMEWORK

National Water Policy

The National Water Policy was first adopted in September 1987. It has been recently reviewed and updated in response to a number of new issues recently have emerged of late. The 2002 National Water policy (Ministry of Water Resources, GOI) has emphasised on several facets that are important from an environmental and social viewpoint including the following statements:

— Water is part of a larger ecological system. Realising the importance and scarcity attached to fresh water, it has to be treated as an essential environment for sustaining all life forms.

— Water is a scarce and precious national resource to be planned, developed and conserved and managed as such, and on an integrated and environmentally sound basis, keeping in view the socio-economic aspects and needs of the States.

— Water resources development and management will have to be planned for a hydrological unit such as drainage basin as a whole or for a sub-basin, multisectored, taking into account surface and ground water for sustainable use incorporating quantity and quality aspects as well as environmental considerations.

— Under para 5 it accords ecology a relatively much lower and fourth priority but indirectly recognises water use for fresh water ecosystems.

— In para 6.3 Project Planning, it states "preservation of the quality of environment and the ecological balance should be a PRIMARY consideration" and goes on to add that the adverse impact on the environment, should be minimised and should be offset by adequate compensatory measures.

— There should be an integrated and multi-disciplinary approach to the planning, formulation, clearance and implementation of projects, including catchments area treatment and management, environmental and ecological aspects, the rehabilitation of affected people and command area development.

— The drainage system should form an integral part of any irrigation project right from planning stage.

— The detrimental environmental consequences of over-exploitation of ground water need to be effectively prevented by the Central and State Governments.

— There should be a close integration of water-use and land-use policies.

— Water allocation in an irrigation system should be done with due regard to equity and social justice.

— Reclamation of water logged/ saline affected land by scientific and cost effective methods should form a part of command area development programme.

— On resettlement and rehabilitation it speaks of the need of a "skeletal national policy" and would like States to evolve their own detailed policies.

— Effluents should be treated to acceptable levels and standards before discharging them in to natural streams and that minimum flow should be ensured in the perennial streams for maintaining ecology and social considerations.

— Special efforts should be made to investigate and formulate projects either in, or for the benefit of, areas inhabited by tribal or other specially disadvantaged groups such as socially weak, scheduled castes and scheduled tribes.

— There are several sections covering aspects of Flood Control and Management, Water Conservation, Drought-prone Area Development, etc., besides Institutional Mechanism, Private Sector Participation and Participatory Approach to water resources management.

National Environment Policy in India

The essential components of environmental policy in India are the prevention of pollution at source; the encouragement, development and application of best available feasible technology; the application of the polluter pays principle; the focus on critically polluted areas, and the participation by the people in the environmental aspects of development. In the Eighth Five-Year Plan, the strategies were set out for prevention and regulation of environmental pollution. For prevention, emphasis was laid on statutory assessment of environmental impacts of projects. For regulation, reliance was placed on the polluter pays principle.

In the Ninth Five-Year Plan, emphasis is being placed on reorienting development policies rather than on direct intervention. It is stated that the measures required to protect the environment will be taken in such a way as to achieve sustainable development. It is further stated that Agenda 21 of the Earth Summit guides the current environmental policies in India. The policy statements bear out the strong commitment that the government has for environmental protection and improvement. In August 2004, the draft National Environment Policy was formulated and placed on the web for wide circulation. The main features that appear new in the proposed policy are shown below:

Definition of environment has been enlarged to include all entities, natural or manmade, external to oneself, which provide value, now or perhaps in the future, to humankind.

Foundational aspirations have been addressed, which are (i) that human beings should be able to enjoy a decent quality of life; (ii) that humanity should become capable of respecting the finiteness of the biosphere; and (iii) that neither the aspiration for the good life, nor the recognition of biophysical limits should preclude the search for greater justice in the world. The principles of good governance, viz., Transparency, Rationality, Accountability, Reduction in time and costs and

Participation with stakeholders, have been incorporated in the action points. Mainstreaming environment in all activities has been explicitly emphasised. Responsibilities of the State and the citizen have been stated in the light of the amendments introducing fundamental duties in the Constitution of India. International commitments on global climate change, stratospheric ozone depletion, and transfer of hazardous wastes and loss of biodiversity have been specifically mentioned.

Certain new principles for environmental management have been included in the policy, some of them based on judicial interpretations. The new principles include the following:

1. The precautionary approach.
2. Polluter pays principle: general and strict liability of the polluter.
3. Economic value of services of environmental resources.
4. Priority for incomparable entities, e.g., Taj Mahal and the tiger.
5. Equity, both intra-generational and inter-generational.
6. Civil liability for environmental damage.
7. State is trustee (not the owner) of all natural resources.
8. Standards should be set in the context of society and economy.

National Forest Policy

The principal aim of Forest Policy is to ensure environmental stability and maintenance of ecological balance including atmospheric equilibrium, which is vital for sustenance of all life forms, human, animal and plant.

The basic objectives that should govern the National Forest Policy are the following:

— Maintenance of environmental stability through preservation and, where necessary, restoration of the ecological balance that has been adversely disturbed by serious depletion of the forests of the country.

— Conserving the natural heritage of the country by preserving the remaining natural forests with the vast variety of flora and fauna, which represent the remarkable biological diversity and genetic resources of the country.

— Checking soil erosion and denudation in the catchment areas of rivers, lakes, reservoirs in the interest of soil and water conservation, for mitigating floods and droughts and for the retardation of siltation of reservoirs.

— Checking the extension of sand-dunes in the desert areas of Rajasthan and along the coastal tracts.

— Increasing substantially the forest/tree cover in the country through massive afforestation and social forestry programmes, especially on all denuded, degraded and unproductive lands.

— Meeting the requirements of fuel-wood, fodder, minor forest produce and small timber of the rural and tribal populations.

— Increasing the productivity of forests to meet essential national needs.

— Encouraging efficient utilisation of forest produce and maximizing substitution of wood.

— Creating a massive people's movement with the involvement of women, for achieving these objectives and to minimize pressure on existing forests.

Essentials of forest management include the following provisions:

- Existing forests and forestlands should be fully protected and-their productivity improved. Forest and vegetal cover should be increased rapidly on hill slopes, in catchment areas of rivers, lakes and reservoirs and ocean shores and, on semi-arid, and desert tracts.
- Diversion of good and productive agricultural lands to forestry should be discouraged in view of the need for increased food production.
- For the conservation of total biological diversity, the network of national parks, sanctuaries, biosphere reserves and other protected areas should be strengthened and extended adequately.
- Provision of sufficient fodder, fuel and pasture, especially in areas adjoining forest, is necessary in order to prevent depletion of forests beyond the sustainable limit. Since fuel wood continues to be the predominant source of energy in rural areas, the programme of afforestation should be intensified with special emphasis on augmenting fuel wood production to meet the requirement of the rural people.
- Minor forest produce provides sustenance to tribal population and to other communities residing in and around the forests. Such produce should be protected, improved and their production enhanced with due regard to generation of employment and income.
- The forest policy provides strategy for area under forests, afforestation, social forestry & farm forestry, management of state forests, rights and concessions, diversion of forest lands for non-forest purposes, wildlife conservation, tribal people and

forests, shifting cultivation, damage to forests from encroachments, fires and grazing, forest-based industries, forest extension, forestry education and forestry research.

ENVIRONMENTAL PROTECTION AND SUSTAINABLE DEVELOPMENT

Environmental Protection and Sustainable Development have been the keystones of the policies and procedures governing the industrial and other developmental activities in India. Ministry of Environment and Forests has taken several policy initiatives and enacted environmental and pollution control legislations to prevent indiscriminate exploitation of natural resources and to promote integration of environmental concerns in developmental projects. One such initiative is the Notification on Environmental Impact Assessment (EIA) of developmental projects issued on 27.1.1994 under the provisions of Environment (Protection) Act, 1986 making EIA mandatory for 29 categories of developmental projects. One more item was added to the list in January, 2000.

The objective of EIA is to foresee and address potential environmental problems/concerns at an early stage of project planning and design. EIA/EMP should assist planners and government authorities in the decision making process by identifying the key impacts/ issues and formulating mitigation measures.

Environmental Clearance of Projects

Any person who desires to undertake any new project in any part of India or the expansion or modernisation of any existing industry or project listed in the Schedule-I of this Notification has to submit an application to the Secretary, Ministry of Environment and Forests, New Delhi.

(a) The application has to be made in the proforma specified in Schedule-II of this notification and has to accompanied by a project report which should, inter alia, include an Environmental Impact Assessment Report, Environment Management Plan and details of public hearing as specified in Schedule-IV prepared in accordance with the guidelines issued by the Central Government in the Ministry of Environment and Forests from time to time. However, Public Hearing is not required in respect of (*i*) small scale industrial undertakings located in (*a*) notified/designated industrial areas/ industrial estates or (*b*) areas earmarked for industries under the jurisdiction of industrial development authorities; (*ii*) widening and strengthening of highways; (*iii*) mining projects (major minerals) with lease area up to twenty five hectares, (*iv*) units located in Export Processing Zones, Special Economic Zones and (*v*) modernisation of existing irrigation projects.

(b) In case of the following site specific projects:

— mining;

— pit-head thermal power stations;

— hydro-power, major irrigation projects and/or their combination including flood control;

— ports and harbours (excluding minor ports),

— prospecting and exploration of major minerals in areas above 500 hectares;

The project authorities have to intimate the location of the project site to the Central Government in the Ministry of Environment and Forests while initiating any investigation and surveys. The Central Government in the Ministry of Environment and Forests then has to convey a decision regarding suitability or otherwise of the proposed site within a maximum period of thirty days. The said site clearance is then granted for a sanctioned capacity and shall be valid for a period of five years for commencing the construction, operation or mining.

c) The reports submitted with the application is evaluated and assessed by the Impact Assessment Agency (IAA), and if deemed necessary it may consult a committee of Experts, having a composition as specified in Schedule-III of the Notification. The Impact Assessment Agency (IAA) is the Union Ministry of Environment and Forests. The Committee of Experts is constituted by the Impact Assessment Agency or such other body under the Central Government authorised by the Impact Assessment Agency in this regard. The said Committee of Experts has full right of entry and inspection of the site or, as the case may be, factory premises at any time prior to, during or after the commencement of the operations relating to the project.

d) The Impact Assessment Agency then prepares a set of recommendations based on technical assessment of documents and data, furnished by the project authorities supplemented by data collected during visits to sites or factories, if undertaken and details of the public hearing.

   The assessment is completed within a period of ninety days from receipt of the requisite documents and data from the project authorities and completion of public hearing and decision conveyed within thirty days thereafter. The clearance granted is valid for a period of five years for commencement of the construction or operation of the project.

ENVIRONMENTAL PROTECTION IN INDIA

Water Pollution

Water quality standards especially those for drinking water are set by the Indian Council of Medical Research. These bear close resemblance to WHO standards. The discharge of industrial effluents is regulated by the Indian Standard Codes and recently, water quality standards for coastal water marine outfalls have also been specified. In addition to the general standards, certain specific standards have been developed for effluent discharges from industries such as, iron and steel, aluminium, pulp and paper, oil refineries, petrochemicals and thermal power plants. Legislation to control water pollution are listed below.

Water (Prevention and Control of Pollution) Act, 1974

This Act represented India's first attempts to comprehensively deal with environmental issues. The Act prohibits the discharge of pollutants into water bodies beyond a given standard, and lays down penalties for non-compliance. The Act was amended in 1988 to conform closely to the provisions of the EPA, 1986. It set up the CPCB (Central Pollution Control Board) which lays down standards for the prevention and control of water pollution. At the State level, the SPCBs (State Pollution Control Board) function under the direction of the CPCB and the state government.

Water (Prevention and Control of Pollution) Cess Act, 1977

This Act provides for a levy and collection of a cess on water consumed by industries and local authorities. It aims at augmenting the resources of the central and state boards for prevention and control of water pollution.

Following this Act, *The Water (Prevention and Control of Pollution) Cess Rules* were formulated in 1978 for defining standards and indications for the kind of and location of meters that every consumer of water is required to install.

Air Pollution

Air (Prevention and Control of Pollution) Act, 1981

To counter the problems associated with air pollution, ambient air quality standards were established, under the 1981 Act. The Act provides means for the control and abatement of air pollution. The Act seeks to combat air pollution by prohibiting the use of polluting fuels and substances, as well as by regulating appliances that give rise to air pollution. Under the Act establishing or operating of any industrial plant in the pollution control area requires consent from state boards. The boards are also expected to test the air in air pollution control areas, inspect pollution control equipment, and manufacturing processes.

National Ambient Air Quality Standards (NAAQS) for major pollutants were notified by the CPCB in April 1994. These are deemed to be levels of air quality necessary with an adequate margin of safety, to protect public health, vegetation and property. The NAAQS prescribe specific standards for industrial, residential, rural and other sensitive areas. Industry-specific emission standards have also been developed for iron and steel plants, cement plants, fertiliser plants, oil refineries and the aluminium industry. The ambient quality standards prescribed in India are similar to those prevailing in many developed and developing countries.

To empower the central and state pollution boards to meet grave emergencies, the *Air (Prevention and Control of Pollution) Amendment Act*, 1987, was enacted. The boards were authorised to take immediate measures to tackle such emergencies and recover the expenses incurred from the offenders. The power to cancel consent for non-fulfilment of the conditions prescribed has also been emphasized in the Air Act Amendment.

The Air (Prevention and Control of Pollution) Rules formulated in 1982, defined the procedures for conducting meetings of the boards, the powers of the presiding officers, decision-making, the quorum; manner in which the records of the meeting were to be set etc. They also prescribed the manner and the purpose of seeking assistance from specialists and the fee to be paid to them.

Complementing the above Acts is the *Atomic Energy Act* of 1982, which was introduced to deal with radioactive waste. In 1988, the *Motor Vehicles Act,* was enacted to regulate vehicular traffic, besides ensuring proper packaging, labelling and transportation of the hazardous wastes. Various aspects of vehicular pollution have also been notified under the EPA of 1986. Mass emission standards were notified in 1990, which were made more stringent in 1996. In 2000 these standards were revised yet again and for the first time separate obligations for vehicle owners, manufacturers and enforcing agencies were stipulated.

Forests and Wildlife

Wildlife (Protection) Act, 1972, Amendment 1991

The WPA (Wildlife Protection Act), 1972, provides for protection to listed species of flora and fauna and establishes a network of ecologically-important protected areas. The WPA empowers the central and state governments to declare any area a wildlife sanctuary, national park or closed area. There is a blanket ban on carrying out any industrial activity inside these protected areas. It provides for authorities to administer and implement the Act; regulate the hunting of wild animals; protect specified plants, sanctuaries, national parks and closed areas; restrict trade or commerce in wild animals or animal articles; and miscellaneous matters.

The Act prohibits hunting of animals except with permission of authorised officer when an animal has become dangerous to human life or property or so disabled or diseased as to be beyond recovery (WWF-India, 1999). The near-total prohibition on hunting was made more effective by the Amendment Act of 1991.

Forest (Conservation) Act, 1980

This Act was adopted to protect and conserve forests. The Act restricts the powers of the state in respect of de-reservation of forests and use of forestland for non-forest purposes (the term 'non-forest purpose' includes clearing any forestland for cultivation of cash crops, plantation crops, horticulture or any purpose other than re-afforestation).

General

Environment (Protection) Act, 1986 (EPA)

This Act is an umbrella legislation designed to provide a framework for the co-ordination of central and state authorities established under the Water (Prevention and Control) Act, 1974 and Air (Prevention and Control) Act, 1981. Under this Act, the central government is empowered to take measures necessary to protect and improve the quality of the environment by setting standards for emissions and discharges; regulating the location of industries; management of hazardous wastes, and protection of public health and welfare. From time to time the central government issues notifications under the EPA for the protection of ecologically-sensitive areas or issues guidelines for matters under the EPA.

Some notifications issued under this Act are:

— *Doon Valley Notification* (1989), which prohibits the setting up of an industry in which the daily consumption of coal/fuel is more than 24 MT (million tonnes) per day in the Doon Valley.

— *Coastal Regulation Zone Notification* (1991), which regulates activities along coastal stretches. As per this notification, dumping ash or any other waste in the CRZ is prohibited. The thermal power plants (only foreshore facilities for transport of raw materials, facilities for intake of cooling water and outfall for discharge of treated waste water/cooling water) require clearance from the MoEF.

— *Dhanu Taluka Notification* (1991), under which the district of Dhanu Taluka has been declared an ecologically fragile region and setting up power plants in its vicinity is prohibited.

— *Revdanda Creek Notification (1989)*, which prohibits setting up industries in the belt around the Revdanda Creek as per the rules laid down in the notification.

— *The Environmental Impact Assessment of Development Projects Notification*, (1994 and as amended in 1997). As per this notification:

  — All projects listed under Schedule I require environmental clearance from the MoEF.

  — Projects under the delicenced category of the New Industrial Policy also require clearance from the MoEF.

  — All developmental projects whether or not under the Schedule I, if located in fragile regions must obtain MoEF clearance.

—

— Industrial projects with investments above Rs 500 million must obtain MoEF clearance and are further required to obtain a LoI (Letter of Intent) from the Ministry of Industry, and an NOC (No Objection Certificate) from the SPCB and the State Forest Department if the location involves forestland. Once the NOC is obtained, the LOI is converted into an industrial licence by the state authority.

— The notification also stipulated procedural requirements for the establishment and operation of new power plants. As per this notification, two-stage clearance for site-specific projects such as pithead thermal power plants and valley projects is required. Site clearance is given in the first stage and final environmental clearance in the second. A public hearing has been made mandatory for projects covered by this notification. This is an important step in providing transparency and a greater role to local communities.

— *Ash Content Notification* (1997), required the use of beneficiated coal with ash content not exceeding 34% with effect from June 2001, (the date later was extended to June 2002). This applies to all thermal plants located beyond one thousand kilometres from the pithead and any thermal plant located in an urban area or, sensitive area irrespective of the distance from the pithead except any pithead power plant.

— *Taj Trapezium Notification* (1998), provided that no power plant could be set up within the geographical limit of the Taj Trapezium assigned by the Taj Trapezium Zone Pollution (Prevention and Control) Authority.

— *Disposal of Fly Ash Notification* (1999) the main objective of which is to conserve the topsoil, protect the environment and prevent the dumping and disposal of fly ash discharged from lignite-based power plants. The salient feature of this notification is that no person within a radius of 50 km from a coal-or lignite-based power plant shall manufacture clay bricks or tiles without mixing at least 25 per cent of ash with soil on a weight-to-weight basis. For the thermal power plants the utilisation of the flyash would be as follows:

— Every coal-or lignite-based power plant shall make available ash for at least ten years from the date of publication of the above notification without any payment or any other consideration, for the purpose of manufacturing ash-based products such as cement, concrete blocks, bricks, panels or any other material or for construction of roads, embankments, dams, dykes or for any other construction activity.

— Every coal or lignite based thermal power plant commissioned subject to environmental clearance conditions stipulating the submission of an action plan for full utilisation of fly ash shall, within a period of nine years from the publication of this notification, phase out the dumping and disposal of fly ash on land in accordance with the plan.

Rules for the Manufacture, Use, Import, Export and Storage of Hazardous Micro-organisms/Genetically Engineered Organisms or Cell were introduced in 1989 with the view to protect the environment, nature and health in connection with gene technology and micro-organisms, under the Environmental Protection Act, 1986. The government in 1991, further decided to institute a national label scheme for environmentally-friendly products called the 'ECOMARK'. The scheme attempts to provide incentives to manufactures and importers to reduce adverse environmental impacts, reward genuine initiatives by companies, and improve the quality of the environment and sustainability of available resources. Besides the above attempts, notifications pertaining *to Recycled Plastics Manufacture and Usage Rules, 1999* were also incorporated under the Environment (Protection) Act of 1986.

The Environment (Protection) Rules, 1986

These rules lay down the procedures for setting standards of emission or discharge of environmental pollutants. The Rules prescribe the parameters for the Central Government, under which it can issue orders of prohibition and restrictions on the location and operation of industries in different areas. The Rules lay down the procedure for taking samples, serving notice, submitting samples for analysis and laboratory reports. The functions of the laboratories are also described under the Rules along with the qualifications of the concerned analysts.

BIBLIOGRAPHY

A. Weintraub, C. Romero, T. Bjørndal, and R. Epstein (Ed), *Handbook of Operations Research in Natural Resources*, Springer, 2007.

Albrecht, S.L., "Equity and Justice in Environmental Decision Making: A Proposed Research Agenda," *Society and Natural Resources*, 1995.

Berrill, N.J., "Regeneration in *Sabella pavonina* (Sav.) and other sabellid worms", *Journal of Experimental Zoology*, 1931.

Berry, Brian L., et. al., *The Social Burdens of Environmental Pollution: A Comparative Metropolitan Data Source*, Cambridge, MA: Ballinger Publishing Co., 1977.

Clarke, Robin, Robert Lamb, Dilys Roe Ward eds.: *Global environment outlook 3 : past, present and future perspectives*, London; Sterling, VA: Nairobi, Kenya : UNEP.

Commission for Africa, "Our common interest: Report of the Commission for Africa", London: Commission for Africa, 2005.

Crandall, D.R., "Postglacial Lahars From Mount Rainier Volcano, Washington", *U.S. Geological Survey Professional Paper*, 1971.

D. Alexander, *Principles of Emergency planning and Management*, Harpended: Terra publishing, 2002.

Dalrymple, G. Brent, *The Age of the Earth*, Stanford: Stanford University Press, 1991.

David L. Hawksworth, "The magnitude of fungal diversity: the 1.5 million species estimate revisited" Mycological Research, Cambridge University Press, 2001.

Divan S and Rosencranz A., *Environmental law and policy in India, cases, materials and statutes,* 2nd edition New York: Oxford University Press, 2001.

GM Wahl, BR de Saint Vincent and ML DeRose, Effect of chromosomal position on amplificationm of transfected genes in animal cells, Nature.

Gupta S., *Country environment review*, policy measures for sustainable development Discussion paper, October 1999 prepared for Asian Development Bank, Programs Department (West) New Delhi: Delhi School of Economics, 1999.

Hiraoka, T., "Nuclear Electricity Generation by Seawater Uranium, *Journal of the Atomic Energy Society of Japan*, 1994.

Hobday, D.K., Offiver, R.A., Parry, G.D., Changes to demersal fish communties in Port Phillip Bay, Australia, over two decades, *Marine and Freshwater Research*, 1999.

Margulis, Lynn; Dorian Sagan, *What is Life?*, New York: Simon & Schuster, 1995.

Meffe, G.K., Nielsen,L.A., Knight R.L., and Schenborn, D.A., *Ecosystem management: adaptive, community-based conservation*, Washington, D.C., U.S.A: Island Press, 2002.

Peters, R. L. and Darling, D. S., "The greenhouse effect and nature reserves", *BioScience*, 1985.

Price, P. W., The resource-based organization of communities, *Biotropica* (Special Issue) 1992.

Proceedings of the National Academy of Sciences, Census of Marine Life (CoML).

Quentin Grafton and Robert Hill, *The Economics of the Environment and Natural Resources*, Blackwell Punlishing, 2004.

Raunkaier, C., *The Life Forms of Plants and Statistical Plant Geography*, Clarendon Press, Oxford, 1934.

Ron Nielsen, *The Little Green Handbook: Seven Trends Shaping the Future of Our Planet*, Picador, New York, 2006.

Saunders, D. A., Hobbs, R. J. and Margules, C. R., "Biology consequences of ecosystem fragmentation: *A review*". *Conservation Biology*, 1991.

Stanley, Steven M., *Earth System History*, New York: W.H. Freeman, 1999.

Sucoff, E., Deforestation, In Environmental Encyclopedia, 2003.

Tjeerd H. van Andel, Eberhard Zangger, Anne Demitrack, "Land Use and Soil Erosion in Prehistoric and Historical Greece' *Journal of Field Archaeology*, Winter, 1990.

Wilson, E. O., "The little things that run the world: The importance and conservation of invertebrates". *Conservation Biology*, 1987.

INDEX

Access and Benefit Sharing Agreements (ABAs) 238
Air pollution 78
Atmospheric pollution 78
Atmospheric pressure 10

Biodiversity protection 252
Biological processes 6
Blue-green bacteria 15

Chaotic system 14
Commercial organic farming 110
Common Agricultural Policy (CAP) 107
Consultative Group on International Agricultural R 230
Continental lithosphere 21
Cultural eutrophication 20
Cyclic agriculture 182

Deforestation 159
Drinking water treatment 281

Environmental problems 1
Environmental systems 2
Eukaryotic cells 32

Food crops 81
Forest catchment areas 167
Forest Stewardship Council (FSC) 160

Genetic diversity 223
Genetic engineering 108
Genetic materials 180
Genetic pollution 234
Genetically modified crops 235
Genetically Modified Organism (GMO) 117
Glacial retreat 19
Global Environment Facility (GEF) 248
Global temperature 6
Green Revolution 106, 234
Greenhouse effect 15
Gross Domestic Product (GDP) 159
Groundwater source 84

Industrial Revolution 268
Intergovernmental Panel on Climate Change (IPCC) 18
Intertropical Convergence Zone (ITCZ) 19
Iron oxides 24

Living Modified Organisms (LMOs) 249

Macroecosystem 33
Metamorphic rocks 23
Microscopic organisms 27

Natural compounds 229
Natural cycles 273
Natural environment 5

Neolithic agriculture 176
Neolithic Revolution 106
Nitrogen fixation 128

Oceanic lithosphere 21
Orbital plane 14
Organochlorine insecticides 123

Paleontological evidence 15
Polycyclic Aromatic Hydrocarbons (PAHs) 284

Reducing Domestic Waste 82
Rock cycle 22

Sedimentary rocks 23
Shrinking forest 171
Soil colour 24
Soil conservation 110
Soil formation 25

Technological developments 107
Terrestrial organisms 28
Traditional conservation 242
Tropical rainforest deforestation 170
Troposphere 9

Ultraviolet radiation 13
United Nations Environment Programme (UNEP) 240
Urban development 170

Water cycle 15

37. Indian Polity **(Revised Edition)**	295/-
38. New Comparative Government **(Revised Edition)**	295/-
Advanced Study in the History of Modern India	
39. (Volume-1: 1707–1813)	225/-
40. (Volume-2: 1813–1920)	325/-
41. (Volume-3: 1920–1947)	175/-
42. Handbook of Nutrition & Dietetics	250/-
43. Development of Education in India	195/-
44. A Textbook of Environment Studies	195/-
45. Teaching of History	175/-
46. Administrative Thinkers	195/-
47. Research Methodology	175/-
48. Curriculum Development	195/-
49. Teaching of Science	175/-
50. Teaching of Mathematics	175/-
51. Principles of Educational & Vocational Guidance	195/-
52. Child Psychology	175/-
53. Abnormal Psychology	175/-
54. Indian Education in Emerging Society	195/-
55. Human Resource Management	175/-
56. Higher Education and Global challenges	175/-
57. Teaching of Geography	175/-
58. Teaching of Social Studies	175/-
59. Teaching of English	175/-
60. PrincipleofOffceManagement(RevisedEdition)	250/-
61. Education for All The Indian Saga	175/-
62. Value Education in Global Perspective	195/-
63. Teacher Training	175/-
64. Public Administration	175/-
65. Public Relations & Integrated Communications	195/-
66. Educational Psychology	175/-
67. Introduction to Educational Technology	175/-
68. Textbook of Food and Nutrition	175/-

Lotus PRESS

Unit No. 220, 2nd Floor, 4735/22,
Prakash Deep Building, Ansari Road, Darya Ganj,
New Delhi - 110002, Ph.: 32903912, 23280047, 09811594448
E-mail: lotus_press@sify.com, www.lotuspress.co.in